T0271152

Urban Resilience and Climate Change in the MENA Region

This book provides an overview of the geopolitical context and climate change risk profile of the Middle East and North Africa (MENA) Region.

Mapping existing scientific literature and key reports on MENA climate change impacts and future projections, Nuha Eltinay and Charles Egbu establish links between the Conference of the Parties (from COP26, COP27 to COP28) Glasgow–Sharm el-Sheikh Work Program for Progress on the Global Goal on Adaptation, and regional climate adaptation financing targets, national government investments, and human security in local case studies. They also address gaps in disaster risk reduction institutional governance for sustainable development in the region. The authors move beyond the existing theoretical understanding of urban resilience to investigate how it is being measured and assessed in MENA in alignment with the IPCC's climate change adaptation indicators. Finally, they explore how disasters and conflict displacement vulnerabilities and fragility affecting the communities most in need are being measured and integrated into cities' resilience action plans and national disaster risk policies.

Providing guidance and policy recommendations based on empirical research and key stakeholder engagement observations, this book will be of great interest to students, scholars, and professionals who are researching and working in the areas of climate change, urban planning, and environmental policy and governance.

As this book comes out just after the closure of The United Nations Climate Change Conference COP28 negotiations, it sets the scene for pre-COP regional context, and paves the way for researchers and practitioners to undertake post-COP28 key takeaways and multi-level government commitments into action, for better climate mitigation and adaptation investments, resilient and sustainable future for all.

Nuha Eltinay is an award-winning spatial planner, a PhD holder in urban resilience from London South Bank University (LSBU), and a master's degree holder in international planning and sustainable development from the University of Westminster, London. In her current role as Senior Expert in Urban Resilience and Climate Adaptation at ICLEI Local Governments for Sustainability (European Secretariat), she leads activities of project management, consultations, and acquisition in sustainable urban development, Ukraine post-war recovery, climate adaptation financing, and urban resilience assessments at cities, local communities and EU regional governments level.

Charles Egbu is Vice Chancellor, Leeds Trinity University. Prior to UEL, he was formerly Pro-Vice Chancellor (Education & Experience) at the University of East London, UK, and former Dean of School of the Built Environment and Architecture at London South Bank University, England, where he also holds the chair in project management and strategic management in construction. Professor Egbu has over 25 years of experience in higher education, has written 12 books, and has lectured nationally and internationally in areas such as sustainable development, resilient communities, construction economics, and innovation and knowledge management in complex environments.

Routledge Focus on Environment and Sustainability

For more information about this series, please visit: www.routledge.com/Routledge-Focus-on-Environment-and-Sustainability/book-series/RFES

Urban Resilience and Climate Change in the MENA Region

Nuha Eltinay and Charles Egbu

Routledge
Taylor & Francis Group

LONDON AND NEW YORK

First published 2024
by Routledge
4 Park Square, Milton Park, Abingdon, Oxon OX14 4RN

and by Routledge
605 Third Avenue, New York, NY 10158

Routledge is an imprint of the Taylor & Francis Group, an informa business

British Library Cataloguing-in-Publication Data
A catalogue record for this book is available from the British Library

ISBN: 978-1-032-42542-9 (hbk)
ISBN: 978-1-032-42543-6 (pbk)
ISBN: 978-1-003-36322-4 (ebk)

DOI: 10.4324/9781003363224

Typeset in Times New Roman
by Apex CoVantage, LLC

Contents

Acknowledgements

First and foremost, thanks and all praises most go to Allah for giving me the wisdom, power, strength, time and support to conduct and complete this book. This publication shall be considered as the most significant research project achieved in my academic and professional career. In completing this book, I am indebted to the following persons and organisations.

I would like to thank my co-author and PhD Supervisor, Professor Charles Egbu, the Vice Chancellor for Leeds Trinity University and previous Dean for the School of Built Environment and Architecture, who patiently and generously provided supervision, advice, support, and guidance on my research during my PhD, resulting in my successfully delivering my thesis, which has built the foundation for this book. Thanks extended to London South Bank University (LSBU) for providing me with the financial support to complete my PhD studies, to my supervisor Professor John Ebohon, the research staff at LSBU who supported the supervision and provided guidance on how to best collect data and apply evidence-based research, as well as external supervisor from Global Disasters Unit at Public Health England (Virginia Murray), who always challenged and inspired me to explore new fields of knowledge.

I am very grateful for the experience gained in my current position as Senior Expert in Urban Resilience and Climate Adaptation at ICLEI Europe and the knowledge captured from individual experts and organisations who were involved in this research as respondents to the interviews and survey questionnaires. This book would not have been completed without them. My greatest gratitude goes to the Arab Urban Development Institute (AUDI), represented by my mentor and AUDI's Director, Mr. Ahmed Al-Salloum, Mark Harvey from Resurgence Urban Resilience Trust and colleagues from the United Nations–Arab States Regional Office for Disaster Risk Reduction, for providing me with access to information and a professional network.

My special thanks go to my husband, Mohamed Rayis, and my treasures, Ahmed, Leen, and Abubakr, for their patience and always bringing the shine to my life. Thanks to my dearest mother Mahasien Hamad, sister Nahla, brothers Nazar and Omer Eltinay, and friends for their great support during the conduction of this long but very productive journey . . . THANK YOU ALL.

Declaration

The authors declare that the work presented in this book, to the best of their knowledge, is original and their own work. Other sources of information used in the study have been well-acknowledged and referenced. Parts of this work have been previously published in the form of presentations, proceedings, or poster format in the following journals, seminars, and conferences.

1 MENA Regional Context – Impact of Climate Change

Background and Geographical Context – Urban Risk Profile

The disastrous impact of climate change on urban livelihoods and natural, bio-diversity systems has long been observed world-wide.

In the first seven years of Sendai Framework for Disaster Risk Reduction (SFDRR) implementation between the years 2015–2021, 145 countries reported a total of 1.05 billion people affected by disasters (Target B). The number affected by disasters per 100,000 per year has nearly doubled, from an average of 1,147 people per year during 2005 to 2014, to 2,066 during 2012 to 2021. In 2021 alone, 38 million new internally displaced people were recorded, of whom over 60 per cent were displaced due to disasters

(GRID, 2022).

In 2022, the Emergency Event Database EM-DAT recorded 387 natural hazards and disasters world-wide that resulted in the loss of 30,704 lives and affected 185 million individuals. Economic losses totalled approximately USD 223.8 billion (CRED, 2023). Shaped by the type of hazard and degree of exposure, extensive disaster risks, derived from urbanisation, environmental degradation, socio-economic inequality, and poor urban governance are witnessed in fragile settings, leading to the accumulation of larger losses in mortality, economic and physical damage (Abu-Awad *et al.*, 2019).

Nowhere else is this so pronounced as in the Middle East and North Africa Region (MENA), where climate change, severity of temperatures, and scarcity of natural resources continue to weaken the absorptive capacity of cities to withstand the impact of natural and man-made hazards. Exacerbated by the prolonged history of civil unrest and protracted displacement, the underlying risks of migration and lack of monitoring of human mobility of refugees and internally displaced persons (IDPs) from camps to urban informal settlements will increase the degree of exposure to climate change, severe weather events, and vulnerability of the urban poor to disasters, leading to "protracted displacements" into capital cities and urban socio-economic centres (Figure 1.1).

DOI: 10.4324/9781003363224-1

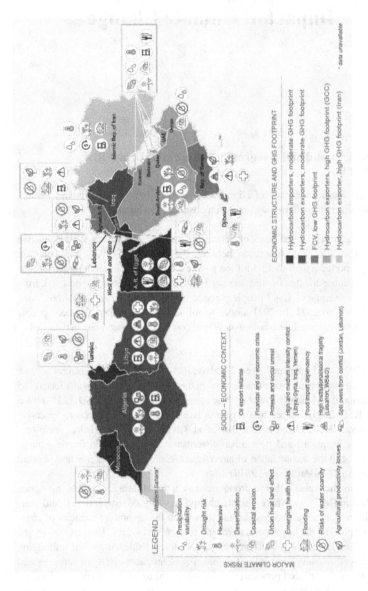

Figure 1.1 Climate change is a threat multiplier in the MENA region (World Bank, 2022)

In the MENA Region, tensions and conflict have erupted as a result of mismanagement, corruption, and the unequal distribution of benefits. Sudden and slow-onset, natural hazards grossly undermine the urban poor displaced by violent conflict, in scale and impact and, in so doing, exacerbate the vulnerabilities of displaced households and generate new patterns of "protracted displacement". This reveals the need to be able to identify the start and the root causes of displacement, in order to gauge its duration (IDMC, 2022). It is also crucial to know when, where, how, and why new and protracted disaster displacements occur, in order to monitor disaster losses and develop mechanisms for building the resilience of IDPs to achieve the 2030 global targets.

Referred to as the "MENA" Region, the Middle East and North Africa Region is divided into four sub-regions, the Mashreq (Eastern) consisting of (Egypt, Iraq, Jordan, Lebanon, Palestine [West Bank and Gaza], Syria, Israel, and Iran), the Maghreb (Western) consisting of (Algeria, Libya, Morocco, Tunisia, and Mauretania), the Gulf Co-operation Council (GCC) countries in the Arabian Peninsula, consisting of (Bahrain, Kuwait, Oman, Qatar, Saudi Arabia, and United Arab Emirates), and the Southern Tier countries (Somalia, Sudan, Comoros, Djibouti, and Yemen). The classification of MENA countries differs between international organisations according to the span of operational networks and geographical scope of activities. For example, the World Bank Report (2014), titled: Natural Disasters in the Middle East and North Africa: A Regional Overview, excluded Sudan and Somalia, while including Djibouti and Malta (World Bank, 2014). According to Majbouri (2015), the listing of countries in the MENA Region extends to include Iran. This also applies to studies by Waha *et al.* (2017) about "climate change impacts in the Middle East and Northern Africa (MENA) region and their implications for vulnerable population groups". According to Choueiri *et al.* (2013), the MENA Region followed the geographical outline of Arab states (Figure 1.2), which is considered in this publication.

Dabbeek and Silva (2019) indicated that less than 15% of the population in the Middle East is at risk from medium and high levels of flooding. However, high concentrations of people and assets in coastal urban centres increase the level of risk and exposure to hazards (UNDRR, 2021). The region is exposed to two major types of natural hazards. Figure 1.3 shows the structure of hazards, types, origins, and effects. The first type of hazard is hydro-metrological hazards, defined as being of "atmospheric, hydrological or oceanographic origin", such as tropical cyclones, floods, drought, heatwaves, cold spells, and coastal storm surges.

These hazards are generated by the North Atlantic Oscillation (NAO), causing storm tracks and annual variations in rainfall in Western and Central North Africa (the Maghreb), most of the Mashreq, and the Arabian Peninsula. In the southern parts of MENA, the Inter-Tropical Convergence Zone (ITCZ) dominates, causing the Indian monsoon system (Donat *et al.*, 2014).

The second type of hazard is geologically generated by the internal processes of the earth in the northern Nubia-Somalia rift zone between Eritrea,

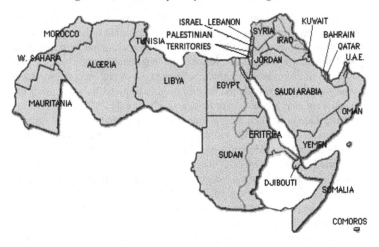

Figure 1.2 Map of Arab States (Choueiri *et al.*, 2013). An overview of the transport sector and road safety in the MENA Region

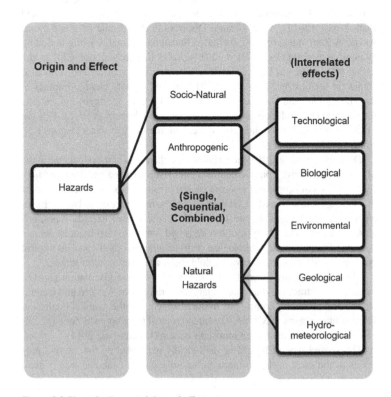

Figure 1.3 Hazards: Types, origin, and effects

Djibouti, Somalia, and Ethiopia. This results in earthquakes and volcanic hazards in this region. Anthropogenic and man-made hazards also exist at lower levels of risk (Poggi *et al.*, 2017). Seismic activity is also a hazard in the Arab region. For example, the Jordan rift valley system places several countries (Jordan, Lebanon, Palestine and Syria) at high risk from earthquakes. Similarly, some countries in the Maghreb Region (Algeria, Morocco and Tunisia) have been exposed to seismic activity in the past. Devastating earthquakes have occurred in Palestine (1927), Lebanon (1956), Morocco (1960), Egypt (1992), and Algeria (2003) (UNDRR, 2013).

While focusing on critically reviewing and analysing extant literature about urban resilience generally, and climate adaptation in particular, the geographical context of the MENA Region is associated in this book with three main bodies of knowledge: resilience, disaster risk reduction (DRR), and sustainability (Carlos *et al.*, 2015), to identify the gap in the breadth and depth of previous research carried out regarding the nexus of disasters, conflict, and displacement in the fragile settings of the Middle East and North Africa Region (Hendrix and Glaser, 2007).

The aim of this book is to develop a new practice intervention in integrating principles of climate adaptation into the process of assessing resilience and gaining support for changing practice intervention in resilience in the legislation of laws regarding disaster. Gaps in the implementation of the voluntary frameworks of the United Nations 2015–2030 are also explored with emphasis on their consistency with assessments of urban resilience at the local level, and the engagement of DRR stakeholders in the decision-making process of the Urban Resilience Action Plan (U-RAP).

Climate Change – Mitigation and Adaptation

Defined as the process of adjustment to actual or expected climate and its effects, the purpose of adaptation in human systems is to moderate or avoid harm or exploit beneficial opportunities. In some natural systems, human intervention can facilitate adjustment to expected climate and its effects (IPCC, 2021). Adaptation to climate change has been recognised as a key measure to reduce the impact of climate change on the most vulnerable. However, this involves costs associated with organising, coordinating, and implementing adaptation, including transaction costs (IPCC, 2007).

With the growing uncertainties regarding climate change scenarios and prediction models caused by extreme and severe weather events and the lack of knowledge of Climate Change Adaptation (CCA) local engineering, operational, and financial realities will increase, widening the gap in financing adaptation (ICLEI, 2011). Multi-lateral development banks (MDBs) and IDFC members who invest in adaptation confront a series of obstacles, including high upfront costs and disbursements before seeing medium- and longer-term gains.

It is noted that despite the substantial nature of upfront public investments and the lack of incentives for the private sector, even when investments adaptation to climate can help to build resilience and generate long-term financial benefits, it remains difficult for local governments to make the case for funding beyond monetary benefits, such as reducing risk to the most vulnerable population or protecting culturally valuable elements, such as historical buildings or recreational spaces. Thus, early action on adaptation is critical to help to realise a "triple dividend" of avoided losses and economic and socio-environmental benefits.

With a high dependency on climate-sensitive agriculture in the MENA Region, investments in Climate Change Adaptation (CCA), critical infrastructure, knowledge strengthening, and policy reforms are essential to help to overcome difficulties with internalising benefits and insufficient incentives to attract private investors within the existing market architecture where a short-term mindset makes investing for the future difficult (World Bank, 2013).

Efforts to reduce or prevent emission of greenhouse gases are also taking place in the MENA Region, in support of national governments reporting on their Nationally Determined Contributions (NDCs) to meet their targets for climate mitigation determined in the Kyoto Protocol. Industrialised MENA countries and economies are in transition to limit and reduce emissions of greenhouse gases (GHGs). Adopted in 1997 and entered into force in 2005, although the Kyoto Protocol binds developed countries to reducing their carbon emissions by an average of 5% below 1990 levels, in the IPCC Sixth Assessment Report, ratified by 192 parties (the European Union, Cook Islands, Niue, and all UN member states except Andorra, Canada, South Sudan, and the United States as of 2022), it is indicated that the "gap between projected emissions based on NDCs in 2030, and emissions pathways compatible with the long-term temperature goal set in the Paris Agreement remains large" (IPCC, 2022).

Ahead of conflict and violence in fragile settings of MENA, countries and national governments set targets to reduce CO_2 emissions and planned for the green transformation of their economies. Nevertheless, the halt of development activities and transfer of finances towards humanitarian response and social security resulted in the destruction of critical infrastructure and deterioration of environmental systems that are required to achieve global and regional targets regarding climate change. This is emphasised further in Chapter 9, addressing progress in the recent COP27 and plans for COP28. With both events taking place in MENA, inter-relations between conflict in MENA and mechanisms for policy development and implementation will be investigated.

Countries in the MENA Region, such as Qatar (hosting the "Doha Amendment" to the Kyoto Protocol for a second commitment period from 2013 to 2020), the Kingdom of Saudi Arabia Paris Agreement pledge stating it aims for 50% of electricity to be generated with renewable energy and 50% with natural gas by 2030 (hosting the MENA Climate Week in Oct 2023), and UAE (hosting the MENA Climate Week in March 2022), demonstrated their commitment to the global climate change agenda. The UAE was announced to be the first country in the MENA Region to launch a plan for net-zero emissions by 2050, and to host the COP28. Egypt, as well, (hosting COP27) demonstrated regional leadership and commitments to take COP26 commitments forward among developing countries at the crossroad between the Arab and African nations.

As this book publication timing comes in the run to COP28, it sets the scene for pre-COP regional context, and paves the way for researchers and practitioners to undertake post-COP28 key takeaways and multi-level government commitments into action, for better climate mitigation and adaptation investments, resilient and sustainable future for all.

Climate Change and Human Security

In the Human Development Report (1994), "Human Security" was defined comprehensively as "Freedom from fear and freedom from wants" and described as safety from threats such as hunger, poverty, disease and environmental threats. The Commission on Human Security defined human security as being:

> to protect the vital core of all human lives in ways that enhance human freedoms and human fulfilment. Human security means protecting fundamental freedoms – freedoms that are the essence of life. It means protecting people from critical (severe) and pervasive (widespread) threats and situations.
>
> (IIHR, 2010)

In the Human Development Report (1994), four fundamental conditions were laid out for human security as being universal, interdependent in its dimensions, people-centred and best secured through prevention. The main goal of human security is to protect the fundamental core of all human lives from basic pervasive threats in a manner that is associated with long-term human satisfaction. Human security integrates the components of security rights and human development.

In relation to the understanding of risk from the DRR perspective, human security involves a comprehensive understanding of threats and causes of insecurity in various dimensions, including economy, food, health, environment, personal, community, and political security (United Nations Trust Fund for Human Security, 2009). In the United Nations Development Programme (UNDP) human security threats in the Arab Region were divided into two categories: hard security and soft security. Hard security includes political instability, foreign intervention, conflicts and threats from vulnerable environmental issues. Soft security includes poverty, hunger, unemployment, lack of health care, and violence. It is important also to define the level of vulnerability and exposure to disasters and risks of conflict, which can vary between urban and rural contexts, affecting the adaptive and transformative capacities required for building resilience and sustaining human security in fragile settings (UNDP, 2009).

Calls for integrating approaches to Climate Change Adaptation (CCA) and Disaster Risk Reduction (DRR) from a human security perspective, are now raised globally more than ever, with the increasing losses caused by disasters and cascading impacts of climate change on the environmental, political, and economic stability of states. The historical development of climate change agendas was aligned with the evolution of the global disaster risk management frameworks. With a shift from managing disasters to managing risks, articulation of the concept of resilience was evident in the progress of the Millennium Development Goals and Climate Change Agreements from the early 1990s to 2015. Noting the slow progress of pre-2015 international agreements in tackling fragility challenges of urban poverty, urban disasters, and urban violence in the Arab Region (Battersby, 2017), global interest in developing better disaster risk management strategies to tackle the root causes of vulnerability emerged. This progress has fuelled the launch of the post-2015 Sustainable Development Agenda and the Sendai Framework for Disaster Risk Reduction, with the aim of both being to reduce disaster damage to critical infrastructure and disruption of basic services (Dora *et al.*, 2015). Followed by the adoption of the COP21 Paris Agreement in the same year, to help limit the increase in temperatures to 2 degrees overall and 1.5°C by the end of the century, global commitments did not stop there.

"While references to the conflict were deleted from the final text, Sendai addresses issues parallel to those that would need to be addressed in prevention and sustaining peace agenda" (Stein & Walch, 2017). The conflict/disaster nexus was strongly outlined by the Conflict Prevention and Peace Forum (CPPF). Founded in 2000 as a programme of the Social Science Research Council as a knowledge broker on the United Nations structural reforms and growing complexity of international conflict and peace operations, in the 2017 publication "The Sendai Framework for Disaster Risk Reduction as a tool for conflict prevention" (Stein & Walch, 2017), three cross-cutting sets of factors were identified that increase both disaster and conflict risks: socio-economic, politico-institutional, and environmental factors. Further support for this argument can be found in a study by Jiuping *et al.* (2016), in which it was indicated that "natural disasters, and particularly climatological disasters, were found to be more likely to trigger longer cumulative social contradictions than any other type of natural disasters" (cited in Theisen *et al.*, 2013). These events often caused local political tensions and even national-level crises, and distracted government attention from the immediate and urgent natural disaster issues (Hendrix *et al.*, 2007; Wagner, 2010, Jiuping *et al.*, 2016).

Underlying risks of climate change can increase the impact of both natural and man-made hazards, threatening human security, and increasing violent conflict in fragile settings. Shaped by discrimination and social exclusion of IDPs and refugees, conflict and violence can also undermine resilience-building at the individual, household, and community level, especially for women and most vulnerable groups (Masson, 2019).

The high cost of displacement and resettlement, induced by poorly managed climate security, extends well beyond the IDPs and refugees directly affected, weakening the absorptive and adaptive capacities of fragile states to reduce risks of disaster. These findings are supported by the views of Kreutz (2010) "that natural disasters could lead to a situation which requires conflict resolution, as governments in an emergency were faced with demands for effective disaster relief, so may need to offer concessions to separatist groups" (cited in

Jiuping *et al.*, 2016). Nevertheless, Renner (2007) drew contrasting views on the opportunities that can arise from large, destructive events to reduce potential conflicts, and establish temporary peace in extremely tense regions. These views are also complemented by the findings of a study by Patel and Nosal (2016), who recognised that existing resilience models "do not acknowledge the possibility or even likelihood of multiple crises", and the views of Renschler *et al.* (2010) regarding resilience in the light of a "more prolonged or multi-dimensional crisis", reflecting the changes across time in the status of fragility and peacebuilding opportunity (cited in Jiuping *et al.*, 2016).

The term migration, as used by Reuveny (2007), provides the choice of relocation, the integration of the concept of human security into forced displacement is suggested to form the term "climate-security displaced people". Climate change displacement is defined by the United Nations Educational, Scientific and Cultural Organisation (UNESCO) as follows:

> [the] displacement of people refers to the forced movement of people from their locality or environment and occupational activities. It is a form of social change caused by a number of factors, the most common being armed conflict. Natural disasters, famine, development, and economic changes may also be a cause of displacement.
>
> (UNESCO, 2017)

Empowering the sense of leadership in building resilience for DRR can reduce the impact of climate change as a transformational driver of forced displacement and increase the effectiveness and efficiency of climate security initiatives. A further action worthy of consideration is to ensure that the challenges and opportunities for building resilience of climate-security displaced people are placed at the centre of disaster recovery and reconstruction, to support inclusive DRR for all community sectors, and ensure the translation of resilience assessment indicators into actions to achieve the Sustainable Development Goals (SDGs).

In addressing the intersection between climate change and human security from the perspective of disaster risk reduction, it is important to consider the 2015–2030 Sendai Framework for Disaster Risk Reduction (SFDRR), the perspective of UN member states on conflict-sensitive economics, displacement, and the geographical and social scope of the impact of climate change (Magnus *et al.*, 2019). In the SFDRR, it is recognised that the state has the primary role to reduce disaster risk and that responsibility should be shared with other stakeholders including local government, the private sector and other stakeholders. Thus, the opportunity is offered in this book to build coherence between the 2015–2030 Sendai Framework for Disaster Risk Reduction (SFDRR) and the 2030 Sustainable Development Goals (SDGs) by integrating human security "hard laws" into to DRR to identify the challenges and priorities for building urban resilience in the MENA Region. This means whether to consider data generated from both quantitative and qualitative methods or resilience assessment indices will continue to adopt numerical methods and remain as "reductionist" techniques.

Reference list

Abu-Awad, B., Abu-Hammad, N., & Abu-Hamatteh, Z. S. H. 2019. Urban and architectural development in amman downtown between natural disasters and great heritage lose: Case study. *International Journal of Architecture and Urban Development*, 9(3): 31–38.

Banerjee, A., Bhavnani, R., Burtonboy, C. H., Hamad, O., Linares-Rivas Barandiaran, A., Safaie, S., Tewari, D., & Zanon, A. 2014. *Natural disasters in the Middle East and North Africa: a regional overview*. World Bank.

Battersby, J. 2017. MDGs to SDGs – New goals, same gaps: The continued absence of urban food security in the post-2015 global development agenda. *African Geographical Review*, 36(1): 115–129.

Carlos, D., Haines, A., Balbus, J., Fletcher, E., Adair-Rohani, H., Alabaster, G., Hossain, R., De Onis, M., Branca, F., & Neira, M. 2015. Indicators linking health and sustainability in the post-2015 development agenda. *The Lancet*, 385(9965): 380–391.

Choueiri, E. M., Choueiri, G. M., & Choueiri, B. M. 2013. An overview of the transport sector and road safety in the MENA region. *Advances in Transportation Studies*, 30.

CRED. 2023. Disasters in numbers. *Centre for Research on the Epidemiology of Disasters*. Available at: https://reliefweb.int/report/world/2022-disasters-numbers

Dabbeek, J., & Silva, V. 2020. Modeling the residential building stock in the Middle East for multi-hazard risk assessment. *Natural Hazards*, 100(2): 781–810.

Donat, M. G., Peterson, T. C., Brunet, M., King, A. D., Almazroui, M., Kolli, R. K., Boucherf, D., *et al.* 2014. Changes in extreme temperature and precipitation in the Arab region: Long-term trends and variability related to ENSO and NAO. *International Journal of Climatology*, 34(3): 581–592.

Dora, C., Haines, A., Balbus, J., Fletcher, E., Adair-Rohani, H., Alabaster, G., Hossain, R., De Onis, M., Branca, F., & Neira, M. 2015. Indicators linking health and sustainability in the post-2015 development agenda. *The Lancet*, 385(9965): 380–391.

Hendrix, C. S., & Glaser, S. M. 2007. Trends and triggers: Climate, climate change and civil conflict in Sub-Saharan Africa. *Political Geography*, 26(6): 695–715.

ICLEI. 2011. Financing the resilient city: A demand driven approach to development, disaster risk reduction and climate adaptation – An ICLEI White Paper. *ICLEI Global Report*. Available at: www.environmental-finance.com/assets/files/Report-Financing_Resilient_City-Final.pdf

IDMC. 2022. Global report on internal displacement 2022. *Internal Displacement Monitoring Centre (IDMC)*. Available at: www.internal-displacement.org/sites/default/files/publications/documents/IDMC_GRID_2022_LR.pdf

IIHR. 2010. *What Is Human Security? Inter-American Institute of Human Rights*. Available at: www.iidh.ed.cr/multic/default_12.aspx?contenidoid=ea75e2b1-9265-4296-9d8c-3391de83fb42&Portal=IIDHSeguridadEN#:~:text=Return%20to%20top-,Concept,are%20the%20essence%20of%20life

IPCC. 2007. *AR4 Climate Change 2007: Synthesis Report. Contribution of Working Groups I, II and III to the Fourth Assessment Report of the Intergovernmental Panel on Climate*. Available at: www.ipcc.ch/report/sixth-assessment-report-working-group-i/

IPCC. 2021. *AR6 Climate Change 2021: The Physical Science Basis. IPCC Sixth Assessment Report, Intergovernmental Panel on Climate Change*. Available at: www.ipcc.ch/report/sixth-assessment-report-working-group-i/

Jiuping, X., Wang, Z., Shen, F., Ouyang, C., & Tu, Y. 2016. Natural disasters and social conflict: A systematic literature review. *International Journal of Disaster Risk Reduction*, 17: 38–48.

Kreutz, J. 2012. From tremors to talks: Do natural disasters produce ripe moments for resolving separatist conflicts? *International Interactions*, 38(4): 482–502.

Magnus, T. O., Gleditsch, N. P., & Buhaug, H. 2013. Is climate change a driver of armed conflict? *Climatic Change*, 117: 613–625.

Majbouri, M, 2015. Calculating the income counterfactual for oil producing countries of the MENA region. *Resources Policy*, 44: 47–56.

Masson, V. L., Benoudji, C., Reyes, S. S., & Bernard, G. 2019. How violence against women and girls undermines resilience to climate risks in Chad. *Disasters*, 43: S245–S270.

Parry, M. L., Canziani, O., Palutikof, J., Van der Linden, P., & Hanson, C. (Eds.). 2007. *Climate Change 2007-Impacts, Adaptation and Vulnerability: Working Group II Contribution to the Fourth Assessment Report of the IPCC*. Vol. 4. Cambridge University Press.

Patel, R., & Nosal, L. 2016. *Defining the resilient city*. New York: United Nations University Centre for Policy Research.

Poggi, V., Durrheim, R., Tuluka, G. M., Weatherill, G., Gee, R., Pagani, M., Nyblade, A., & Delvaux, D. 2017. Assessing seismic hazard of the East African Rift: A pilot study from GEM and AfricaArray. *Bulletin of Earthquake Engineering*, 15: 4499–4529.

Pörtner, H. O., Roberts, D. C., Poloczanska, E. S., Mintenbeck, K., Tignor, M., Alegría, A., Craig, M., Langsdorf, S., Löschke, S., Möller, V., & Okem, A. 2022. IPCC, 2022: Summary for policymakers.

Renner, M. 2007. *Beyond disasters: creating opportunities for peace* (Vol. 173). Worldwatch Institute.

Renschler, C. S., Frazier, A. E., Arendt, L. A., Cimellaro, G. P., Reinhorn, A. M., & Bruneau, M. 2010. *A framework for defining and measuring resilience at the community scale: The PEOPLES resilience framework* (pp. 10–0006). Buffalo: MCEER.

Reuveny, R. 2007. Climate change-induced migration and violent conflict. *Political Geography*, 26(6): 656–673.

Stein, S., & Walch, C. 2017. The Sendai framework for disaster risk reduction as a tool for conflict prevention. In: *Conflict Prevention and Peace Forum*.

Theisen, O. M., Gleditsch, N. P., & Buhaug, H. 2013. Is climate change a driver of armed conflict?. *Climatic change*, 117: 613–625.

Tol, R. S. J., & Wagner, S. 2010. Climate change and violent conflict in Europe over the last millennium. *Climatic Change*, 99: 65–79.

UNDP, 1994. *Human Development Report*. Oxford University Press. New York.

UNDP, 2009. *Challenges to Human Security in the Arab Countries*. New York.

UNDRR, 2013. Factsheet: Overview of Disaster Risk Reduction in the Arab Region.

UNDRR. 2021. Overview of disaster risk in the Arab region. *2021 Regional Assessment Report for the Arab States*. Available at: www.undrr.org/2021-regional-assessment-report-arab-states

UNESCO. 2017. *Climate change displacement*. United Nations Educational, Scientific and Cultural Organisation.

Waha, K., Krummenauer, L., Adams, S., Aich, V., Baarsch, F., Coumou, D., Fader, M., Hoff, H., Jobbins, G., Marcus, R., & Mengel, M., 2017. Climate change impacts in the Middle East and Northern Africa (MENA) region and their implications for vulnerable population groups. *Regional Environmental Change*, 17: 1623–1638.

World Bank. 2013. *Adaptation to Climate Change in the Middle East and North Africa Region*. Available at: http://web.worldbank.org/archive/website01418/WEB/0__C-152.HTM

World Bank. 2022. Middle East and North Africa Roadmap (2021–2025). *Driving Transformational Climate Action and Green Recovery in MENA*.

Xu, J., Wang, Z., Shen, F., Ouyang, C., & Tu, Y. 2016. Natural disasters and social conflict: A systematic literature review. *International journal of disaster risk reduction*, 17: 38–48.

2   Drivers of Climate Risk – Vulnerabilities

Urbanisation of Poverty

In the MENA Region, "it is difficult to generalise urbanisation trends because the region comprises a great diversity of socio-economic, human, natural resources and characteristics" (Madbouly, 2009). Underlying urbanisation drivers of poverty and weak governance also shape the patterns of land tenure security, highly affecting the livelihoods of farmers and pastoral communities.

One in five people in MENA already live within 60 km of conflict and the region is home to a quarter of forcibly displaced people in the world (16.3 million in 2016). Being the world's most water-scarce region, and with 60% of people living in high or extremely high water-stressed areas, the challenges of water management and food security are considered from the perspective of urban governance and decentralisation at the local level. Exposed to multiple risks and interlocking crises of urban poverty, urban violence, and urban disaster, the concept of fragile city urbanisation is explored in the context of the MENA Region, taking into account the gap in the implementation of regional DRR policies and global frameworks adopted at the local level (Lienard, 2022).

In the Middle East and North Africa (MENA) Region, millions of people live in coastal areas, representing approximately 53% of the region's total population (World Bank, 2011). An increase in the MENA population up to 436,720 million was reported in the recent 2016 World Bank Development Indicators (World Bank, 2017). Further, "in North Africa, there could be up to 19.3 million internal climate migrants by 2050, accounting for 9% of the sub-regions total projected population" (World Bank, 2022).

The lack of adequate and functional infrastructure and services, accompanied by the unplanned urbanisation, will result in an increase in the level of exposure and vulnerability to natural hazards, mostly affecting the urban poor (UN Habitat, 2012). In 1980, the urban areas of the MENA Region accounted for 48% of the total population and, by 2000, this had increased to 60%. The estimated 1990–2003 average growth rate of 2.1% per annum was registered, and urban share of total population growth from 48% was significant against an average of 54% for all developing countries. In the UN 2020 projections, with an estimation of rapid increase over the past ten years, the MENA

DOI: 10.4324/9781003363224-2

population was expected to increase to "430 million, of which 280 million are expected to be urban" (World Bank, 2008). This increase in population in the MENA Region has resulted in a pattern of "urbanisation of poverty", with "cities in poor countries now exceeding the total population in cities of the industrialised countries" (Piel, 1997).

It is also important to differentiate between the definitions of "urban" and "rural". Evidence from primary data collection indicated the need to have a clear understanding of terminologies to monitor internal displacement patterns better. An expert from the Internal Displacement Monitoring Centre begins with an injunction, follows that with a reflection from an experience in differentiating between rural and urban settings in monitoring data about disaster loss and displacement, and then offers additional commentary on that issue.

> First of all, we think it's important to be really careful about the data we're providing. From the IDMC point of view, we don't support statements that say that 60% to 80% of the people that are internally displaced are in urban areas, because we don't have data disaggregated by urban, rural, and what do we mean by urban resilience or by rural resilience? The lack of rural resilience pushes you to an urban area. Urban risk and urban disaster risk pushes you even secondarily, to go back to rural areas. So, I mean, there are other complexities there that we would like to lay out. We are getting more and more interested about displacement in urban areas.
>
> (IDMC, 2018)

Supported by the definition of "urban resilience" by Meerow *et al.* (2016) as

> the ability of an urban system and all its constituent socio-ecological and socio-technical networks across temporal and spatial scales to maintain or rapidly return to desired functions in the face of a disturbance, to adapt to change, and to quickly transform systems that limit current or future adaptive capacity"

the complexity in defining what is "urban" is noted based on the study of 25 publications by Meerow *et al.* (2016).

The categorisation of "urban" varied in the literature between referring to cities as "complex systems" (Brugmann, 2012; Cruz *et al.*, 2013; Da Silva *et al.*, 2012), composed of networks, or a combination of both (Desouza and Flanery, 2013; Godschalk, 2003, Cutter *et al.*, 2016). Concerns regarding the vagueness in defining "urban" was also raised by IDMC, indicating the growing urban risk of displacement associated with the dynamics of intra-city movements:

> Urban displacement can be of different forums; it can be urban to urban within the city, you can even have short, forced movements.

The problem with urban is that there's not a definition of urban. The joint research centre of the European Union came up with this map that explains what they mean by urban. It's more focused on the built environment, the assets and the people that live within that environment. But knowing that urban and the relation between the urban and the pre-urban and rural is very dynamic and very contextual, thus we need a standard definition on what, in IDMC, we mean by urban, at least for our monitoring purpose.

(IDMC, 2018)

Cutter *et al.* (2016) applied non-parametric rank analysis, analysis of variance, and logistic regression in a study to help to describe the relationships between rurality and disaster resilience in contrast to resilience in urban areas. This study can be used a guide for identifying urban-rural differences in resilience to disaster by pinpointing the driving factors or characteristics of resilience in rural America compared with metropolitan America. Nevertheless, this remains constrained by contextual barriers, and opportunities for generalisation might not apply (Cutter *et al.*, 2016). These theoretical principles were supported with primary data collected from interviews with an IDMC representative who indicated that:

Resilience applies in both disaster risk reduction and conflict. But it depends a lot on the context, meaning urban or rural, also we need to have gender distinction because a man or a woman in an urban Middle Eastern city is not like in Latin America. So, I think that the contextual definition of resilience is very important especially for global frameworks and global metrics because we cannot standardise resilience for everyone everywhere. It also has a series of cultural aspects behind it.

(IDMC, 2018)

Climate change, severe environmental conditions and desertification, and complex, customary land rights dominate the rural settings, in association with the deprived dispute resolution systems and the lack of legal recognition that can lead to conflict and generate prolonged patterns of protracted displacement. As noted by Zimmermann (2011):

progress is mainly technology-driven (e.g., the geo-industry) and too often not accompanied by progress in reforming land policies, improving the normative framework, involving civil society, and re-engineering institutional processes.

According to the Global Report on Internal Displacement, "disasters triggered by natural hazards caused twice as many new displacements in 2015 as conflict and violence" (cited in Glasser, 2016). This report brings attention to the significance of conflict and violence as drivers of protracted "climate-change-induced displacement" that is considered in this case for both internally displaced persons (IDPs) and refugees (Parry *et al.*, 2014; Reuveny, 2007). Exposed to multiple risks and inter-locking crises of urban poverty, urban violence, and urban disaster, the lack of monitoring of the human mobility of refugees and IDPs from camps to urban informal settlements will worsen their living conditions and social exclusion with deprived infrastructural services and lack of spatial planning policies that are discussed in Chapter 8.

This will increase their degree of exposure to climate change, severe weather events, and vulnerability to disasters, leading to protracted displacements into capital cities and urban socio-economic centres. Noting that climate change will push approximately 132 million people into poverty by 2030 (World Bank, 2022), mostly affecting small farmers and fishery industries, climate change can serve as "push" factor in migration, jeopardising the capacity of already populated urban centres, and can lead to social unrest, already unfolding in the context of instability in the region's most fragile settings (Figure 2.1).

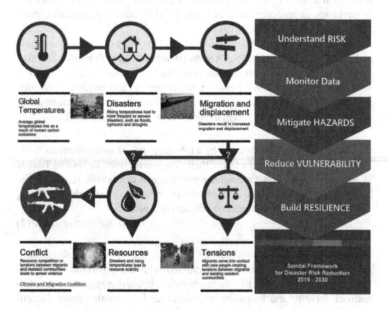

Figure 2.1 Exploring evidence of the climate change and conflict connection (Climate and Migration Coalition, 2015)

Decentralisation and Urban Governance

Urban governance of land tenure includes "traditional practices for making decisions on land transactions, inheritance, resettlement and the resolution of land disputes beyond formal institutions and government authorities" (Mitchell, 2011). Decentralisation in the MENA Region is confined to the political devolution of authority of central agencies to the local level, causing an increase in the autonomy of local government in planning and decision-making, with limited resources for urban management and service delivery (World Bank, 2008). As stated by Madbouly (2009), "the limited fiscal transfers and human resources at hand, and the limited financial and political autonomy severely impede local government capacity to finance, deliver and manage urban services" (cited in Madbouly, 2009). With the lack of accountability and transparency, the growing demands for infrastructure and affordable housing, the high cost of land in the informal sector, and poor public land management all resulted in the proliferation of slums and informal settlements.

"Decentralisation is still viewed mainly as an administrative technique (amounting to deconcentration rather than devolution) and not as a political process" (Bergh, 2010). With the increase in the power of supervision by centrally-appointed government representatives, the "geography of discontent", and limited local financial autonomy, the implementation of legal frameworks at the local level and the development of self-revenue-generation mechanisms is increasing dependency on already short-falling central government transfers and subsidies (OECD, 2019).

Democratisation at the national level has long been observed by analysts and policymakers in the MENA Region (Bergh, 2010), with limited attention to the intersection between urbanisation, decentralisation, local governance, and the need for tackling the root causes of conflict and political instability, such as poverty, limited access to services, and employment opportunity, that was witnessed strongly in recent events such as the "Arab Spring".

> The Arab Spring is proving to be a turning point for some countries in the MENA Region. After the fall of their previous regimes, Egypt and Tunisia are now engaged in far-reaching political and democratic reform, while Morocco and Jordan are also introducing reforms aimed at decentralising political powers and increasing political competition.
>
> (O'Sullivan *et al.*, 2012)

The breakdown of state governance in the region's war-affected states, such as Syria, Libya, and Yemen, is combined with the economic and social losses inflicted by conflict in these countries, impacting regional and international security and humanitarian, social, and economic affairs. Decentralisation contributes to reforming administrative structures, empowering local governments to deliver policies and public services needed locally, while improving local self-governance and bringing power and decision-making competences to the most vulnerable people (OECD, 2019). Associated with the growing demands of IDPs and refugees for tenure security and access to services, it helps to provide greater efficiency in areas such as health care, infrastructure, education, and land reforms, which can also serve host communities and provide social equity. Nevertheless, this cannot be achieved without the devolution of powers, the independence of local administrations, the adoption of new legislation to help to enhance transparency, and civil society involvement in the implementation of reforms. Post-war and conflict recovery situations in the fragile settings of MENA can create an opportunity to embed such measures into new urban governance systems and open the opportunity for better integration of DRR policies and urban resilience strategies into the structural reforms of governments.

Noting the impact of climate change on the degradation of natural resources that can result in conflict over food security and access to natural resources, decentralisation can help to enhance natural resource-based employment and livelihoods. Restoring governance in degraded lands and access to water after natural disasters can operate as an opportunity for conflict resolution. Taking into account the role that humanitarian aid can play in this regard, it is important to direct funds towards integrating decentralisation and urban governance reforms into the overall programming, strategic planning, and governance of the funds to bridge the gap between emergency response and long-term development, while maintain the sustainability and impact of interventions beyond the short-term scales of projects.

Fragility and Political Instability

Defined by Robert Muggah (2015) as "discrete metropolitan units whose governance arrangements exhibit a declining ability and/or willingness to deliver on the social contract" (Muggah & Savage, 2012), the inter-linkages between internal displacement and city fragility are recognised in the IDMC 2014 Global Overview as not only having "the largest displaced populations in the region, but the Fund for Peace also ranks them among the world's top five fragile states" (IDMC, 2015). This was reflected by the United Nations University Centre for Policy Research where "in some cities, systems of law and order, ranging from the police, judiciary, penal systems and other forms of legal enforcement, are dysfunctional and considered illegitimate by the citizens who they are intended to serve". In order to investigate this further, John de Boer (2015) developed a conceptual framework of three main components that shape the fragile city: "urban disasters", "urban poverty", and "urban violence".

These components were explored in the historical post-colonial context of fragility in the MENA Region Arab States, and its impact on framing the chronic vulnerability of IDPs and refugees to forced evictions and relocation to achieve sustainable and resilient "durable solutions". The In the United

Nations Economic and Social Commission for Western Asia (ESCWA) 2017 Report, it is indicated that 40% of Arab countries are immersed in or have lived through armed conflict in the past six years.

> The region's historical geopolitical instability and impact of climate change severe and frequent events continue to undermine the Arab states' adaptive capacity to shocks and stresses, and shifting central governments' investments from disaster risk reduction into ad-hoc emergency response.
>
> (ESCWA, 2017)

Aligned with the increase of a series of anti-government protests, uprisings and armed rebellions that spread across the MENA Region in late 2010 in recognition of the Arab Spring, these movements triggered civil wars in the years that followed in Syria (2011), Iraq (2014), and Yemen (2015). Figure (2.2) shows an outline of conflicts in the Arab Region by type for the years 1946–2018 (ESCWA, 2018).

The number of refugees hosted by Arab countries rose from 7 million to almost 8.5 million. While the region has only 5.4% of the world's population, it hosts 37.5% of its refugees. In association with displacement, key findings from the Global Report on Internal Displacement (GRID, 2018) show that the region accounted for only 38% of the global total of 11.8 million internally displaced people within their own countries. Divided into displacements

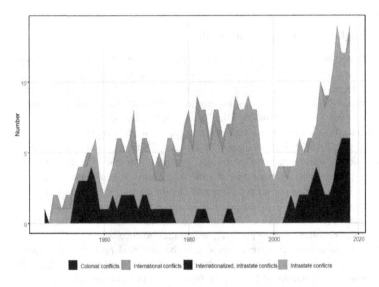

Figure 2.2 Conflicts in the Arab Region by type, 1946–2018 (ESCWA, 2019)

caused by disaster and conflict, it is noted that the Middle East and North Africa (MENA) Region has a relatively low figure of internal displacement because of disasters in comparison with South Asia, East Asia, Pacific, and the Americas, reaching approximately 232,000. On the contrary, numbers of internal displacement in the MENA rose to 4.5 million for people who fled within their own countries to escape conflict and violence in 2017.

Considering how disaster risk reduction could positively influence peace-building and conflict resolution based on the findings of a systematic literature review about natural disasters and social conflict, Jiuping *et al.* (2016) indicated that "large destructive natural disasters sometimes provided opportunities to reduce existing and potential conflicts, and to establish a temporary peace, particularly in extremely tense regions". Noting that disasters are not all natural, a historical review of violent conflicts was applied in a study by Billon and Waizenegger (2007) in order to understand the underlying drivers of displacement vulnerability and gain insight into the root causes of fragility in the region. This will help to associate inter-linkages in monitoring between intensive and extensive risks in fragile settings and how the implementation of the SFDRR can work as a peacebuilding method in the context of the Arab Region. Dividing the historical timeline of violent conflict into the pre-and-post colonial period, investigating the definition of the term "post-colonial" further in the literature review was considered, as addressed by the Foreign Policy Research Institute in Table 2.1. This will help to associate inter-link-ages with the development of DRR Policy in the region's modern history and provide evidence for how this term is applied in this book.

Table 2.1 Definition of the term "post-colonial" – The Foreign Policy Research Institute

Category	Post-Colonial States and the Struggle for Identity in the Middle East since World War II
Temporal Category	• The Middle East States that were controlled by European and Ottoman Empires in the 19th and early 20th centuries. • Throughout the 20th century, colonial rule crumbled and various post-colonial states emerged.
Type of State and Type of Politics	• The states that emerged from colonial empires in the Middle East and inherited the colonial institutions that were in place and designed by the colonial powers to control the populations from above. • Non-democratic institutions led by military officers who had served in armies that were run by the British, the French, and the Ottomans and continued to perform that function after the fall of colonial empires. • These institutions were designed to protect the state from its own people rather than to protect it from outside militaries.

(Continued)

Table 2.1 (Continued)

Category	Post-Colonial States and the Struggle for Identity in the Middle East since World War II
	• These states have been susceptible to coups, which often came in waves of 30 years following World War II (Syria four coups, Iraq three coups) and often compounded the problems of weak states and strong societies.
Post-Modern Critique	• Referral to the pillars of modernity (liberalism, free markets, secularism, etc.) as not being rational concepts that resulted from logical or scientific deliberation but, rather, are social constructions that developed out of a particular Western, often Christian, experience. Modernity, therefore, is Western, and imperial powers have imposed the Western ways of thought associated with it in the Middle East.

The history of conflict in the post-colonial era in MENA's Arab States started with the Israeli occupation and the 1947–1949 Palestine war where approximately 750,000 Palestinians fled or were expelled from their homes to seek refuge in neighbouring Jordan, Syria, and Lebanon. The roots of that modern conflict between Jews and Arabs and the precursor to the Arab–Israeli conflict are dated back to the year 1881 when approximately 565,000 Arabs and 24,000 Jews lived in Palestine. Nevertheless, violent conflict was officially triggered by the British-led intervention for the United Nations Special Committee on Palestine (UNSCOP) to divide Palestine into two states. The Arab state comprised approximately 42% of Palestine, and the Jewish State approximately 55% of the remaining territory, including Jerusalem as an international zone. This was followed by the United States interventions during the period 1979–1996, including the 1979 Iran hostage rescue effort, deployments to Lebanon in 1982, and the 1990 Gulf War.

The United States-led occupation of Iraq has led to massive regional suffering, increased regional instability, and created an Iraqi refugee crisis. Continuing operations in northern and southern Iraq in the wake of the defeat of Baghdad and international tensions with Turkey, Iran, Russia and other global powers, threatened the escalation of new forms of conflict, and recently shaped the politics of the Syrian War. The Carnegie Middle East Centre provides a wider perspective of the conflict and fragility in Arab States beyond the MENA boundaries, with emphasis on the impact of the Israeli occupation on Palestine, Lebanon, Syria, Jordan, and Egypt, while "stretching the definition of the region, one might also include the 1992 peacekeeping operation in Somalia" (Lesser *et al.*, 1998). Internal divisions have led to civil war in Iraq, Lebanon, and Yemen, and quasi-civil war in Algeria and Palestine, and Somalia has collapsed (Salem, 2010).

Taking into account the brief historical overview outlined above, civil wars and internal disputes causing protracted displacement within countries are not highlighted in existing literature, hence this book provides insight into the state of post-colonial experience of MENA countries regarding the highest number of three coups across the region, as well as hosting the highest number of refugees for decades. Considering the type of state fragility and type of political criteria applied by the Foreign Policy Research Institute, it is worth noting that despite the severe losses caused by conflict, urban stresses caused by climate risk deepen socio-economic vulnerabilities. As stated by Harries *et al.* (2013): "the number of high profile disasters in fragile and conflict-affected states have increased the attention being paid to how disasters and conflict collide, through systematic analysis is limited and sometimes contested" (cited in Basanta, 2014).

Thus, the risk of displacement because of disaster should not be overlooked, as the cumulative effects of climate change, rapid urbanisation, water scarcity, environmental degradation, socio-economic inequity, and violent conflict have emerged as main drivers of the increasing vulnerabilities to disasters in fragile states (Parry, 2004). In the MENA Region, monitoring the human mobility of refugees and IDPs remains a necessity, while integrating their coping capacities and resilience to the risk of disaster into the national plans of hosting countries to reduce the risk of disaster, to frame joint coordination mechanisms between conflict and DRM stakeholders, financial investments of international humanitarian response, national DRR policy legislations, and to raise awareness and holistic understanding of compound risk factors and vulnerability to natural and man-made hazards in fragile and conflictual settings.

Reference list

Basanta, L., & Alfonso, J. 2014. *Disaster Risk Reduction Contribution to Peacebuilding Programmes*. Uppsala Universitet, Sweden.

Bergh, S. 2010. Decentralisation and local governance in the MENA region. In *IEMed Mediterranean Yearbook*, pp. 253–258. Mediterranean Commission of the UCLG (United Cities and Local Governments), Local and Regional Authorities in the New Mediterranean Governance, Marseille.

Billon, P. L., & Waizenegger, A. 2007. Peace in the wake of disaster? Secessionist conflicts and the 2004 Indian Ocean tsunami. *Transactions of the Institute of British Geographers*, 32(3): 411–427.

Brugmann, J. 2012. Locating the 'Local Agenda': preserving public interest in the evolving urban world. In *Scaling urban environmental challenges* (pp. 348–371). Routledge.

Climate Change and Migration Coalition. 2015. *Climate change, armed violence and conflict. Infographic: exploring evidence for the climate change and conflict connection.* United Kingdom.

Cruz, S. S., Costa, J. P. T., de Sousa, S. Á., & Pinho, P. 2013. Urban resilience and spatial dynamics. *Resilience thinking in urban planning*, 53–69.

Cutter, S. L., Ash, K. D., & Emrich, C. T. 2016. Urban – rural differences in disaster resilience. *Annals of the American Association of Geographers*, 106(6): 1236–1252.

Da Silva, J., Kernaghan, S., & Luque, A. 2012. A systems approach to meeting the challenges of urban climate change. *International Journal of Urban Sustainable Development*, 4(2), 125–145.

De Boer, J. 2015. Resilience and the fragile city. *Stability: International Journal of Security and Development*, 4(1).

Desouza, K. C., & Flanery, T. H. 2013. Designing, planning, and managing resilient cities: A conceptual framework. *Cities*, 35: 89–99.

ESCWA. 2017. Situation Report on International Migration: Migration in the Arab Region and the 2030 Agenda for Sustainable Development. United Nations Economic and Social Commission for Western Asia.

ESCWA. 2018. *Implementation of the Istanbul Programme of Action for the Arab Least Developed Countries for the Decade 2011 to 2020*. Productive Capacity Progress and Challenges in Mauritania, Sudan and Yemen. United Nations Economic and Social Commission for Western Asia.

ESCWA. 2019. *Trends and Impacts in Conflict Settings. No. 6. Developing a Risk-Assessment Framework for the Arab Region*. United Nations Economic and Social Commission for Western Asia.

Expert from the Internal Displacement Monitoring Centre. 2018. (Interview, 23 March).

Glasser, R. 2016. *Can Sendai Framework Help Conflict Prevention*? United Nations Office for the Coordination of Humanitarian Affairs. Available online: https://reliefweb.int/report/world/can-sendai-framework-help-conflict-prevention

Godschalk, D. R. 2003. Urban hazard mitigation: Creating resilient cities. *Natural hazards review*, 4(3): 136–143.

Harris, K., Keen, D., & Mitchell, T. 2013. When disasters and conflicts collide: Improving links between disaster resilience and conflict prevention. ODI.

IDMC. 2018. *Off the GRID. Making Progress in Reducing Internal Displacement*. Norwegian Refugee Council.

IDMC Global Overview. 2015. *People Internally Displaced by Conflict and Violence*. Norwegian Refugee Council.

Jiuping, X., Wang, Z., Shen, F., Ouyang, C., & Tu, Y. 2016. Natural disasters and social conflict: A systematic literature review. *International Journal of Disaster Risk Reduction*, 17: 38–48.

Lesser, I. O., Nardulli, B. R., & Arghavan, L. A. 1998. Sources of conflict in the greater Middle East. *Sources of conflict in the 21st century: Regional futures and US strategy*, 171–229.

Lienard, C. 2022. *MENA Climate Week 2022: Tackling Climate Change in MENA by Improving Regional Cooperation. Rethinking Security in the 2020s Series – Policy Brief*. Brussels International Center.

Madbouly, M. 2009. Revisiting urban planning in the Middle East North Africa region. *Regional Study Prepared for UN-Habitat Global Report on Human Settlements*. Available Online: http://www.unhabitat.org/grhs/2009

Meerow, S., Newell, J. P., & Stults, M. 2016. Defining urban resilience: A review. *Landscape and Urban Planning*, 147: 38–49.

Mitchell, D., & Garibay, A. 2011. *Assessing and Responding to Land Tenure Issues in Disaster Risk Management*. Food and Agriculture Organisation of the United Nations (FAO).

Muggah, R., & Savage, K. 2012. Urban violence and humanitarian action: Engaging the fragile city. *The Journal of Humanitarian Assistance*, 19(1): 2012.

Organisation for Economic Co-operation and Development. 2019. *Making Decentralisation Work: A Handbook for Policy-Makers*. OECD Publishing.

O'Sullivan, A., Rey, M.-E., & Mendez, J. G. 2011. Opportunities and challenges in the MENA region. *Arab World Competitiveness Report 2012*, pp. 42–67. Available Online: http://www.oecd.org/mena/49036903.pdf

Parry, M. L., Rosenzweig, C., Iglesias, A., Livermore, M., & Fischer, G. 2004. Effects of climate change on global food production under SRES emissions and socio-economic scenarios. *Global Environmental Change*, 14(1): 53–67.

Piel, G. 1997. The urbanisation of poverty worldwide. *Challenge*, 40(1): 58–68.

Reuveny, R. 2007. Climate change-induced migration and violent conflict. *Political Geography*, 26(6): 656–673.

Salem, P. 2010. *The Arab State: Assisting or Obstructing Development?* Carnegie Endowment for International Peace.

Un-Habitat. 2012. *The State of Arab Cities Report, Challenges of Urban Transition. State of Cities – Regional Reports*. Routledge.

World Bank. 2008. *Urban Development in MENA*. Available at: http://web.worldbank.org/archive/website01418/WEB/0__CO-51.HTM

World Bank. 2011. *Poor Places, Thriving People: How the Middle East and North Africa Can Rise Above Spatial Disparities*. The World Bank.

World Bank. 2017. *Data Bank: World Development Indicators*. Available at: https://databank.worldbank.org/source/world-development-indicators

World Bank. 2022. *Middle East and North Africa Roadmap (2021–2025). Driving Transformational Climate Action and Green Recovery in MENA*. Available Online: https://www.worldbank.org/en/region/mena/publication/middle-east-north-africa-climate-roadmap

Zimmermann, W. 2011. Towards land governance in the Middle East and North Africa region. *Land Tenure Journal*, 1.

3   Urban Resilience –
Conceptual Framework

Definition of Resilience

Defined by the United Nations Office for Disaster Risk Reduction (2016) as

> the ability of a system, community or society exposed to hazards to resist, absorb, accommodate to, and recover from, the effects of a hazard in a timely and efficient manner, including through the preservation and restoration of its essential basic structures and functions.

the term "resilience" is explored broadly and varies between literature about different disciplines such as engineering, psychology, and disaster, yet all agree on the role of the social dynamics of local communities in understanding the scale of risk, level of vulnerability, and strengthening governance of urban resilience.

Holling (1973) wrote about the bio-physical system of resilience in a publication about resilience and stability of ecological systems, defining resilience as "determining the persistence of relationships within a system and a measure of the ability of these systems to absorb changes of state variable". Adger (2000) expanded the concept to build the resilience of a social system based on the social capital of communities and institutions, and the resilience of the ecological systems on which they depend. In a publication titled "Social and ecological resilience: Are they related?", Adger defined resilience "in the broadest sense of habitualized behaviour and rules and norms that govern society, as well as the more usual notion of formal institutions with memberships, constituencies and stakeholders". This school of thought is adopted strongly in the philosophical position of this book and guides the investigation of the roles and responsibilities of key stakeholders in DRR in Chapter 7. This is also supported by Shepherd and Dissart (2022), who defined resilience based on a perception of capability, "which fosters the agency people require to withstand and adapt to change, and is related to a human security notion of development where justice and equality play a critical role", and its inter-relationship with the concept of climate security.

Re-defining resilience in association with fragility and vulnerability to climate change in the context of the MENA Region will help to bridge the

DOI: 10.4324/9781003363224-3

gap between DRR local and national platforms, complementing the matrix of local assessments of resilience with the sustainability of durable solutions to displacement from the perspective of Climate Change Adaptation (CCA) while identifying inter-links between the Nansen Initiative Protection Agenda for Disaster Displacement and the principles of the IASC Framework for durable solutions for internally displaced persons (Chapter 6). This approach is framed around extending the phenomenon of "climate refugees" into "climate security displaced" (CSD) people, moving beyond the complexity of defining the status of refugees by the international community to demonstrate how the intersection between disaster management policy and conflict fragility can affect human mobility patterns of internal displacement and beyond.

Moving towards combining the social-ecological and environmental systems, this concept has evolved further to be focused on the capacity to adapt to, shape, and change, and on how the societal dimension can enable management based on ecosystem CCA. In a publication titled "Adaptive governance of social-ecological systems", Folke *et al.* (2005) provided a conceptual basis for measuring resilience by developing a disaster resilience of place (DROP) model. This model represents resilience as "a dynamic process dependent on antecedent conditions, the disaster's severity, time between hazard events, and influences from exogenous factors", which is embedded in the understanding of this study of the complexity of the vulnerability of CSD people in environments of disasters and conflict and the dimension of time in shaping the parameters of protracted displacement (cited in Cutter *et al.*, 2008).

Further understanding of the definition of resilience was explored using the evidence generated from a literature review of 172 publications by Meerow *et al.* The literature about urban resilience over a period of 41 years was identified using Elsevier's Scopus and Thompson Reuters Web of Science on Holling's Socio-Ecological System Framework (Walker *et al.*, 2004). Table 3.1 shows the definitions identified in the areas of risk management, hazards, adaptation to climate change, and sustainability as the key areas of investigation in this book.

Table 3.1 Urban Resilience: Definitions (Meerow *et al.*, 2016)

Focus Areas	Literature	Definition
Risk Management	(Rose, 2007) Economic resilience to natural and man-made disasters: Multi-disciplinary origins and contextual dimensions (cited in Coaffee, 2008).	The ability of an entity or system to maintain functionality (e.g., continue producing) when shocked (Rose, 2007). It describes the ability of nation-states and government agencies to develop disaster mitigation processes and "hardened" critical national infrastructure to ensure that they can continue operating within the global economy at their regular capacity.

Focus Areas	Literature	Definition
Hazards	(UNDRR, 2016) Terminology on Disaster Risk Reduction. Geneva (cited in Gaillard, 2010).	The ability of a system, community, or society exposed to hazards to resist, absorb, accommodate, and recover from the effects of a hazard in a timely and efficient manner, including through the preservation and restoration of its essential basic structures and functions (UNDRR, 2016).
Climate Change Adaptation	(IPCC, 2007; Tyler and Moench, 2012; Adger, 2000).	The ability of a social or ecological system to absorb disturbances while retaining the same basic structure and ways of functioning. The capacity of self-organisation and the capacity to adapt to stress and change (IPCC, 2007).
Sustainability	(Mileti, 1999) Disasters by design: A re-assessment of natural hazards in the United States, Natural hazards and disasters (cited in Cutter *et al.*, 2008; Ollenburger and Tobin, 1999).	The ability to tolerate – and overcome – damage, diminished productivity, and reduced quality of life caused by an extreme event without significant outside assistance (Mileti, 1999).

In Table 3.1, there is a relationship between the four definitions in defining the location where resilience occurs, varying between an entity or a system. However, the UNDRR definition highlighted the social impact of resilience by adding the terms "community" and "society", stressing the impact of CSD people on weakening the social cohesion and solidarity of local communities, while signifying the transitional points between "community" and "society" in building resilience, as indicated by an expert in displacement studies from the Department of Architecture at the University of Khartoum:

The relocation of IDPs is a transitional process from the community scale (solidarity, support and of waiver of needs for others) to a wider society scale (each member has an individual role and responsibility). This movement has its own triggers that DRR governmental resilience-building strategies should feed into, and work towards building a mature community – Communityness is the basic element for building resilience.

(UoK, 2014)

Acting as an educational hub to support community development, the University of Khartoum (UoK) is a multi-campus, co-educational, public university, considered the largest and oldest university in Sudan. UoK was founded as Gordon Memorial College in 1902 and was officially established in 1956 when Sudan gained independence from British colonisation. Contributing to peacebuilding and knowledge-sharing, the UoK community-based research and partnerships with local and international role-players provided a broader and a holistic assimilation of the understanding of resilience. Taking into account the recent conflict and urban violence that escalated in Sudan between the Sudanese Military and Rapid Forces in April 2023, it is important to highlight the impact of these events on the regional security and stability of MENA and the complexity of the political situation driven by the continuous attempts of the government to suppress the interests of ethnically-based, regional movements and youth-led, local governance. This has led to the prolonged marginalisation and social exclusion of IDPs and extended the socio-economic gap between community and society, which first appeared in the Darfur conflict 20 years ago, into the capital, Khartoum, and triggered urban violence across its localities.

Resilience of What and to What?

The 2014 Guidelines for Resilience Systems Analysis, published by the Organisation for Economic Co-operation and Development (OECD), were adhered to in this book, to investigate the questions of resilience of what and to what to help analyse risk and develop a set of directions to build resilience to climate change and conflict. Further investigation was applied to the UNDRR definition of resilience and its effectiveness in achieving sustainability, where "system" is divided into socio-economic, environmental, and institutional components to explore the means of collecting data, analysing, and using outcomes of data analytics in developing an Urban Resilience Action Plan (U-RAP) for human settlements that is investigated further in Chapter 5.

Risk can be "cumulative and compounding gradually until a tipping point is reached and transformed into a shock" (World Bank, 2015). Moving from defining risks to understanding the drivers behind risk, the findings of a study by Lischer (2007) showed that "in the migration literature, conflict is only one of many causes of displacement. Other causes of migration include environmental degradation, natural disasters, and economic incentives". These concepts are explored further in the following sections to identify the parameters required to build urban resilience in the context of the MENA Region.

Evidence to date suggests that there is a gap between social vulnerability and bio-physical vulnerability indices. The findings of a study by Cutter (2003) about social vulnerability indicated that the relationship between the two types should be investigated further as a "complex interaction" to help to advance understanding of vulnerability on a local, regional, and national scale. Accordingly, this publication moved beyond the "positivist epistemology for urban resilience research, arguing that phenomena can be objectively defined and measured" (Miller *et al.*, 2010). The research approach was dominated by a constructivist and interpretivist position in questioning "resilience of what" and "resilience to what" to investigate the associations between climate security-induced displacement (independent/exploratory) variables against resilience (dependent/response) corresponding variables of exposure, vulnerability, and absorptive capacities.

Addressing the unanswered questions of "resilience of what" and "resilience to what" will guide the investigation of climate change mitigation and adaptation in this chapter. Being inter-related, each aspect has a consequential impact on the other, affecting the socio-economic, technological, physical, institutional, and behavioural principles of urban resilience.

Data accessibility, analytics, and useability are also explored for "hazards" and categorised into "shocks" and "stresses" in association with the time scale of impact. The guiding principles in a paper titled "Defining Disaster Resilience: A DFID Approach" published by the Department for International Development (DFID) in 2011 also helped to shape the epistemological understanding of the terminologies "shocks" and "stresses", building on the levels of resilience determined in the DFID Framework shown in Figure 3.1. Adopting the working definition of DFID for "disaster resilience" as "the ability of countries, communities and households to manage change, by maintaining or transforming living standards in the face of shocks or stresses without compromising their long-term prospects" (DFID, 2011), hazards fall into the categorisation of "shocks" as a sub-category for "disturbance" in responding to the question "Resilience to what?"

"Shocks" are sudden events that can cause a state of disaster-related shock (disease outbreaks, weather-related and geophysical events, floods, high winds, landslides, droughts or earthquakes), or a state of conflict-related shock (outbreaks of fighting, violence, or shocks related to economic volatility) (DFID, 2011). "Stresses" are longer-term trends that undermine the performance of a given system and increase the vulnerability of role-players within it. Stresses can act as underlying drivers of risk such as natural resource degradation, loss of agricultural production, urbanisation, demographic changes, climate change, political instability, and economic decline (DFID, 2011). It is noted that in the DFID framework for resilience (Figure 3.1), the vertical relationship between shocks and stresses is also outlined, where the response curve to shocks and stresses could be gradual and uneven because of the political context, secondary shocks, or lack of information. This is essential to consider in the fragile context of the MENA Region because of the economic and political pressures placed by migration, urbanisation, and demographic changes on the fragile urban systems of cities and the lack of basic services (Mirkin, 2010).

In addressing the question "Resilience of what?", it is important to consider the actions of resistance and absorption associated with the ability of the physical environment to withstand the damage imposed by hazards, referred to as "resistance" (Norton & Chantry, 1993). With regard to resilience, using the term "resistance" is limited to the physical components of a system and ignores the socio-ecological context. Absorbance is associated with the economic context of resilience, as cited in the Knowledge Platform for Disaster

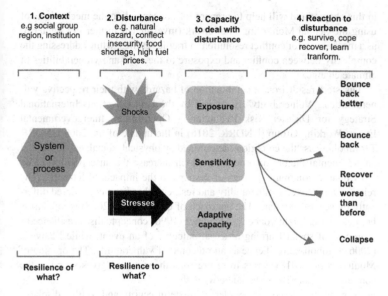

Figure 3.1 Four Element Resilience Framework (DFID, 2011)

Risk Reduction. A "country's resilience depends, to an important extent, on whether the government's institutional system can absorb financial losses" (UNDRR, 2013). This can be adopted to accommodate the context of the fragile cities of the MENA Region by utilising existing absorptive, adaptive, and transformative capacities. The boundaries of what can be listed under the terminologies of "timely" and "efficient", noted in the UNDRR definition, cannot be measured in the context of small-scale and slow-onset disasters, which can greatly undermine the long-term efforts of resilience. The measures of disruption and quality of response can also vary according to local emergency response protocols and national DRR policies, which vary depending on DRR governance systems and cannot be generalised to fit a specific time frame (Eltinay and Harvey, 2019).

Dynamics of Resilience (Exposure, Risk, and Vulnerability)

The dynamics of resilience are explored in this chapter by investigating the theoretical underpinning of "exposure", "risk", and "vulnerability", which serves to frame inclusion and exclusion criteria for exploring contradictions, gaps, and inconsistencies in the literature, and understanding how risk governance and humanitarian action operate in the region. An investigation of the gaps in implementing regional DRR policies and agreements is undertaken in this chapter and will help to prepare for understanding the mechanisms of using the Sendai Monitoring Framework (investigated further in Chapter 6) as a mechanism for conflict resolution in fragile settings, while addressing the complexity between conflict and exposure to the risks and vulnerabilities of climate change.

Disasters result from a combination of hazards with their respective vulnerabilities. "Vulnerability" is defined by the United Nations International Strategy for Disaster Risk Reduction – Open-Ended Intergovernmental Expert Working Group (UNDRR, 2016) in the update of the 2009 UNDRR Terminology as "the conditions determined by physical, social, economic and environmental factors or processes, which increase the susceptibility of an individual, a community, assets or systems to the impacts of hazards". The relationship between vulnerability and resilience has been engendered differently by several scholars. The approach of Cutter *et al.* (2008) is to address both as two different concepts. Handmer (1996) conceptualises resilience as a process of social learning or as an outcome of an event, while Manyena (2006) symbolises resilience as an outcome of vulnerability. The opinion of Modica *et al.* (2018) varies from previous theories and gives scope to the common characteristics shared between the two.

Moving beyond country borders, climate insecurity and conflict displacement forces people to flee their countries and seek refuge in neighbouring states. The United States Department of Defense reported that: "Climate change can act as a threat multiplier for instability in some of the most volatile regions of the world" (Nordås and Gleditsch, 2007). Evidence from a systematic literature review of published literature in which social conflicts resulting from natural disasters were examined for the period 1986 to 2013, indicated that "natural disaster caused social conflicts and intensified social disorder and instability resulting from the disaster, greatly increased the risk of social crisis and commonly undermined government coping strategies" (Endfield *et al.*, 2004; Fisher, 2010, Jiuping *et al.*, 2016). Integrating the human security perspective into DRR policies and programming will help to increase the visibility and awareness of conflict sensitivity in building resilience for climate security. Accordingly, the concept of risks of social conflict arising from natural disasters was brought into this study based on the views of Jiuping *et al.* (2016), as shown in Figure 3.2, to determine the correlation between natural disasters and disaster-based conflicts through a systematic literature review of research which was focused on social conflicts caused by geo-spatial disasters between the period of 1986 to 2013.

Combining both refugees and IDPs, an overview of the historical evolution of conflict and instability in the MENA Region was applied for the post-colonial period to date, reflecting the CSD concept. Noting that the influx of Syrian refugees since 2011 was regarded as "the world's unprecedented scale of forced migration since second world war" (ESCWA, 2018), this phenomenon is investigated in association with environmental and socio-political

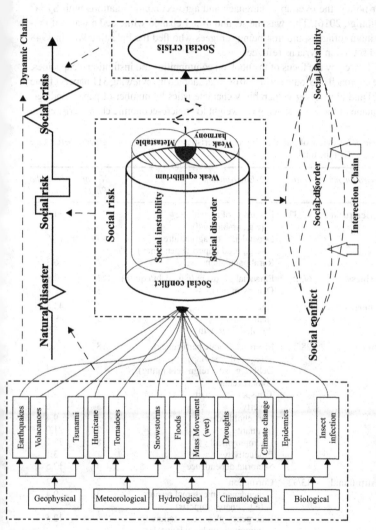

Figure 3.2 The Dynamic interaction system for social conflict caused by natural disasters (Jiuping *et al.*, 2016)

degradation of climate security caused by droughts and by triggering violent conflict between the Syrian opposition and government forces. Lebanon shared the highest number of Syrian refugees fleeing into the neighbouring countries of Jordan and Turkey "owing to its geographic proximity, open border policy, the overlap in language and agro-economic relations with Syria" (Dionigi, 2016). This was the second time Lebanon witnessed a wave of Palestinian refugees, the first being refugees who fled the Palestine War in 1948 and settled in Syria as refugees.

Driven by the focus of the book on communities and institutional capacities, the approach of Modica *et al.* (2018) was adopted in reviewing 311 papers (Table 3.2) and identifying vulnerability characteristics by number of papers and percentage of the total, which gave weight to "macro-economic characteristics" in

Table 3.2 Review of vulnerability characteristics by number of papers and percentage of the total

Environment	Total	Sub-Environment	No. of Papers	% of the total
Agricultural	12/32	Extension of agriculture (e.g., arable land)	11	34.4
		Dependency on agriculture (e.g., food import dependency)	5	15.6
		Rural population	2	6.3
Business	6/32	Financial exposure (e.g., debt/equity)	1	3.13
		Density of business	3	9.4
Demographic	16/32	Age	14	43.8
		Gender	3	9.4
		Population growth	2	6.3
Economic	28/32	Macro-economic performance (e.g., GDP saving)	18	56.3
		Debt (e.g., sovereign debt rating)	3	9.4
		Total revenue	2	6.3
		Transportation costs	1	3.1
		Poverty	13	40.6
		Household debt	3	9.4
		Inequality	7	21.9
		Unemployment	8	25
		Productivity	1	3.1
		Sectorial dependence	4	12.5
Institutional	13/32	Corruption	2	6.3
		Dependence on external resource (e.g., energy imports)	3	9.4
		Emergency plans (e.g., failure to communicate knowledge)	4	12.5
		Government effectiveness (e.g., governance index)	2	6.3
		Institutional capacity	6	18.8
		Political rights	5	15.6

(Continued)

Table 3.2 (Continued)

Environment	Total	Sub-Environment	No. of Papers	% of the total
Land	18/32	Land use (e.g., relative urban entropy)	3	9.4
		Population pressure (crowding)	13	40.6
		Urbanisation (e.g., formation of slums)	5	15.6
Material	8/32	Infrastructure characteristics (e.g., road density)	3	9.4
		Building characteristics (e.g., number of buildings)	8	25
Natural	10/32	Air pollution	2	6.3
		Ecosystem conversion (e.g., % land unmanaged)	4	12.5
		Ecosystem service value	1	3.13
		Environmental sustainability	2	6.3
		Erosion	2	6.3
		Soil pollution	2	6.3
		Water pollution	5	15.65
Risk	11/32	Insurance	1	3.13
		Population at risk	4	12.5
		Previous disaster effects (e.g., number of people affected)	6	18.8

both concepts, derived from the theory that "socio-economic conditions influence both the inherent characteristics of individuals, community, and network infrastructures". On closer inspection, it is worth considering literature about "vulnerability" as a stand-alone focus on the variables of measuring community capacities, while investigation of the concept of "resilience" in isolation favours institutional capacities (Modica *et al.*, 2018).

In this chapter, the theoretical foundation was established for extending the phenomenon of "climate refugees" into "climate security displaced" (CSD) people, which is investigated further in this book, moving beyond the complexity of defining the status of refugees by the international community to demonstrate how the intersection between disaster management policy and conflict fragility can affect human mobility patterns of internal displacement. Taking into account the theory of resilience as being "the ability of households, communities, and nations to absorb and recover from shocks, whilst positively adapting and transforming their structures and means for living in the face of long-term stresses, change and uncertainty" (Harris *et al.*, 2013), the definition by UNDRR of resilience can be advanced by strengthening three different types of capacities: absorptive, adaptive, and transformative capacities. Responding to the questions "Resilience to what?" and "Resilience of what?", these concepts are investigated further in Chapter 6, exploring how the inter-relationship between these capacities over time can be associated with the quality of datasets available for preparedness and pre-disaster mitigation of climate change while further guiding the strategies required to reduce the impact of post-disaster risks and losses (Béné *et al.*, 2012; Papadopoulos *et al.*, 2017).

Reference list

Academic and Researcher from the University of Khartoum, Sudan. 2014. (Interview, 17 July).

Adger, W. N. 2000. Social and ecological resilience: Are they related? *Progress in Human Geography*, 24(3): 347–364.

Adger, W. N. 2006. Vulnerability. *Global Environmental Change*, 16(3): 268–281.

Béné, C., Wood, R. G., Newsham, A., & Davies, M. 2012. Resilience: New utopia or new tyranny? Reflection about the potentials and limits of the concept of resilience in relation to vulnerability reduction programmes. *IDS Working Papers, No. 405*, pp. 1–61. Available Online: http://dx.doi.org/10.1111/j.2040-0209.2012.00405.x

Coaffee, J. 2008. Risk, resilience, and environmentally sustainable cities. *Energy Policy*, 36(12): 4633–4638.

Cutter, S. L. 2003. The vulnerability of science and the science of vulnerability. *Annals of the Association of American Geographers*, 93(1): 1–12.

Cutter, S. L., Barnes, L., Berry, M., Burton, C., Evans, E., Tate, E., & Webb, J. 2008. A place-based model for understanding community resilience to natural disasters. *Global Environmental Change*, 18(4): 598–606.

DFID. 2011. *Defining Disaster. Resilience. A DFID Approach Paper*, pp. 1–20. London.

Dionigi, F. 2016. The Syrian refugee crisis in Lebanon: State fragility and social resilience.

Eltinay, N., & Harvey, M. 2019. Building urban resilience in the Arab region: Implementing the Sendai Framework for Disaster Risk Reduction 2015–2030 at the local level. *Input Paper for the Global Assessment Report*. Available online: http://www.undrr.org/publication/building-urban-resilience-arab-region-implementing-sendai-framework-disaster-risk

Endfield, G. H., Fernández Tejedo, I., & O'Hara, S. L. 2004. Conflict and cooperation: Water, floods, and social response in colonial Guanajuato, Mexico. *Environmental History*, 9(2): 221–247.

ESCWA. 2018. *Implementation of the Istanbul Programme of Action for the Arab Least Developed Countries for the Decade 2011 to 2020*. United Nations Economic and Social Commission for Western Asia.

Fisher, S. (2010). Violence against women and natural disasters: Findings from post-tsunami Sri Lanka. *Violence against women*, 16(8): 902–918.

Folke, C., Hahn, T., Olsson, P., & Norberg, J. 2005. Adaptive governance of social-ecological systems. *Annual Review of Environment and Resources*, 30: 441–473.

Gaillard, J.-C. 2010. Vulnerability, capacity and resilience: Perspectives for climate and development policy. *Journal of International Development: The Journal of the Development Studies Association*, 22(2): 218–232.

Handmer, J. 1996. Policy design and local attributes for flood hazard management. *Journal of Contingencies and Crisis Management*, 4(4): 189–197.

Holling, C. S. 1973. Resilience and stability of ecological systems. *Annual Review of Ecology and Systematics*, 4(1): 1–23.

IPCC, 2007. Climate change 2007: Appendix to synthesis report. Climate change 2007: Synthesis report. Contribution of working groups I, II and III to the fourth assessment report of the intergovernmental panel on climate change, 76–89.

Jiuping, X., Wang, Z., Shen, F., Ouyang, C., & Tu, Y. 2016. Natural disasters and social conflict: A systematic literature review. *International Journal of Disaster Risk Reduction*, 17: 38–48.

Lischer, S. K. 2007. Causes and consequences of conflict-induced displacement. *Civil Wars*, 9(2): 142–155.

Meerow, S., Newell, J. P., & Stults, M., 2016. Defining urban resilience: A review. *Landscape and urban planning*, 147: 38–49.

Mileti, D. 1999. *Disasters by design: A reassessment of natural hazards in the United States*. Joseph Henry Press.

Miller, F., Osbahr, H., Boyd, E., Thomalla, F., Bharwani, S., Ziervogel, G., Walker, B., et al. 2010. Resilience and vulnerability: Complementary or conflicting concepts? *Ecology and Society*, 15(3).

Mirkin, B. 2010. *Population Levels, Trends and Policies in the Arab Region: Challenges and Opportunities*. United Nations Development Programme, Regional Bureau for Arab States.

Modica, M., Reggiani, A., & Nijkamp, P. 2018. Vulnerability, resilience and exposure: Methodological aspects and an empirical application to shocks. *SEEDS Working Paper 13*.

Nordås, R., & Gleditsch, N. P. 2007. Climate Change and Conflict, *Political Geography*, 26(6): 627–638.

Norton, J., & Chantry, G. 1993. Promoting principles for better typhoon resistance in buildings – A case study in Vietnam. In: *Natural Disasters: Protecting Vulnerable Communities: Proceedings of the Conference Held in London, 13–15 October 1993*, pp. 533–546. Thomas Telford Publishing.

Ollenburger, J. C., & Tobin, G. A. 1999. Women, aging, and post-disaster stress: Risk factors. *International Journal of Mass Emergencies & Disasters*, 17(1): 65–78.

Papadopoulos, T., Gunasekaran, A., Dubey, R., Altay, N., Childe, S. J., & Fosso-Wamba, S. 2017. The role of big data in explaining disaster resilience in supply chains for sustainability. *Journal of Cleaner Production*, 142: 1108–1118.

Rose, A., 2007. Economic resilience to natural and man-made disasters: Multidisciplinary origins and contextual dimensions. *Environmental Hazards*, 7(4): 383–398.

Shepherd, P. M., & Dissart, J. 2022. Reframing vulnerability and resilience to climate change through the lens of capability generation. *Ecological Economics*, 201: 107556.

Tyler, S., & Moench, M. 2012. A framework for urban climate resilience. *Climate and development*, 4(4): 311–326.

UNDRR (2013). *Factsheet: Overview of Disaster Risk Reduction in the Arab Region*. Available from: https://www.unisdr.org/we/inform/publications/31693

UNDRR. 2016. *Terminology on Disaster Risk Reduction*. United Nations Office for Disaster Risk Reduction. Available at: www.preventionweb.net/terminology/view/477

Walker, B., Holling, C. S., Carpenter, S. R., & Kinzig, A. 2004. Resilience, adaptability and transformability in social – ecological systems. *Ecology and Society*, 9(2).

World Bank (2015). *City Strength Diagnostic Methodological Guidebook*. First Edition. Washington.

4 Measuring Urban Resilience

Assessments of Urban Resilience

The theory developed by Walker and Salt (2012) is applied in this chapter to guide the approach used in this book for measuring and building resilience, by understanding that:

> resilience is not a single number or a result. It is an emergent property that applies in different ways and in the different domains that make up your system. It is contextual and it depends on which part of the system you are looking at and what questions you are asking.
>
> (cited in Quinlan *et al.*, 2016)

Thus, in this study, the globally standardised definition agreed by the UNDRR is re-defined to fit the fragile context of the MENA Region (Walker & Salt, 2012).

Following the variation in definitions of the concept of resilience indicated in Chapter 3, the debate about the assessment of resilience is continued to help to direct investments into Climate Change Adaptation (CCA) and programmatic pillars for budgeting towards sustainable development with the adoption of indicators of resilience as key components of measuring the success of long-term programmes for CCA (Bahadur *et al.*, 2015; Schipper & Langston, 2015) in line with the prominence of the concept of resilience in development discourse after 2015 (FSIN, 2016).

There is no shortage of literature reviews and comparative studies about resilience frameworks, yet there are no agreed principles on criteria for comparing indicators "because of the tremendous discrepancies in conceptual frameworks and aims of the different indicator sets, coupled with the numerous existing perspectives on resilience" (Schipper & Langston, 2015). Being strongly influenced by global policies and diverse conceptual entry points regarding resilience, scholars addressed the similarities and differences between resilience frameworks from different perspectives.

This was reflected strongly in the work developed by Quinlan *et al.* (2016) in which a summary of approaches was framed to measure and assess

DOI: 10.4324/9781003363224-4

resilience in association with the contexts for definitions of resilience that emerged in the literature. While acknowledging the principles of socio-eco-logical resilience shared between the different frameworks, clear guidance was not provided in this analytical study for inclusion of the ten resilience frameworks compared. Nevertheless, the study included guidance for the development of a conceptual model of an integrated, social-ecological sys-tem where "people and ecosystems interact as an integrated social-ecological system, representing an important advance in sustainability science more broadly, and is a foundational concept for resilience assessment" (Berkes and Folke, 1998, Quinlan *et al.*, 2016).

An important contribution offered in this section is to highlight the differences in defining resilience by different agencies operating on the ground identified by an expert in urban studies from the Overseas Development Institute (ODI):

> What you have in Urban Spaces is a diversity of resilience understanding, as we need to separate between resilience as a buzzword and resilience at the operational level. Derived by organisations like the Rockefeller Foun-dation, resilience is defined differently, from donors' perspective it comes in general terms, while from a humanitarian perspective, how you link relief and development and how to operate in urban spaces is different in comparison to camps management and control. On the other hand, extend-ing drivers of the resilience agenda has to be operationalized in the SDGs, to support local governments in the implementation of the Sustainable Development Goals.
>
> (ODI, 2018)

Assessment of resilience involves a process of identifying how resilience is created, maintained, or broken down. A primary objective is to conceptualise a place and associated issue(s) by focusing on system dynamics to compare various future pathways and to identify those that are robust to shocks and other drivers of change (Walker *et al.*, 2002). The main challenge in breaking down resilience-building components is to translate the results and indicators of assessing resilience into action plans.

Similarly, an analytical review of 43 resilience frameworks by Bahadur *et al.* (2015) was applied to "examine the degree to which resilience frame-works align with conceptual understandings of 'resilience thinking' to assess their internal coherence and rigour" (Schipper & Langston, 2015). By iden-tifying seven key issues and entry-points of frameworks, most "frameworks consider resilience in the context of disasters and DRR (23), followed by climate change adaptation (11), then food security (10), livelihoods (10), poverty (2), ecosystems (2), and conflict (2)" (Schipper and Langston, 2015). Taking into account the context of fragility and urban violence in the MENA Region, these findings led to including in the scope of this study the finding of analytical frameworks for resilience that place more emphasis on the fragile context of the region and address the vulnerabilities of IDPs and refugees. In view of this, a background paper by the United Nations University Centre for Policy was featured in this study because it identified 35 frameworks for measuring fragility, risk, and resilience (Bosetti *et al.*, 2016). The analytical study brought attention to applicability at the city level commonly targeted by the frameworks assessing resilience. Bosetti *et al.* (2016) emphasised the gap between national and local platforms as "fragility continues to be analysed predominantly from a state-centric perspective and pays little attention to the sub-national dimensions of fragility". Although relatively comprehensive, the context of displacement remains missing.

Aimed at fostering harmonisation of the approaches and methods available to help cities to assess their strengths, vulnerabilities, and exposure to a multitude of natural and man-made threats", the techniques deployed by previous scholars were compared with the 36 methods outlined in a literature review by Sharifi (2016) to identify criteria for the selection of methods in view of the shift of the global community from "disaster management" to "disaster reduction" with the initiation of the Hyogo Framework for Action and the adoption of the 2015–2030 Sendai Framework for Disaster Risk Reduction, building coherence with the Sustainable Development Goals and the HABITAT III New Urban Agenda. Accordingly, in this book, developing more theoretically-based conceptual entry points is avoided in favour of a more practical contribution to the materialisation of the 2015–2030 global targets.

Measuring Urban Resilience

An important contribution offered in this book is the differentiation between the "assessment" and "measurement" of resilience. In the process of understanding why and how DRR stakeholders measure resilience, Kerner and Thomas (2013) indicated that "managers must monitor and measure what they manage". Nevertheless, in emergent studies and resilience theories, the gaps and challenges of measuring and assessing complex concepts of resilience using a narrow set of indicators are highlighted. This argument is defended with the view that "reducing resilience to a single unit of measurement may block the deeper understanding of system dynamics needed to apply resilience thinking and inform management actions" (Quinlan *et al.*, 2016).

The format of an assessment can vary depending on the expert-driven, participatory approach and the level of detail sought, as indicated in the 2010 Resilience Assessment Guide that listed thresholds and tipping points, adaptive cycles of change, cross-scale interactions, and adaptive governance (Quinlan *et al.*, 2016). Falling into the category of building urban resilience in fragile settings, this understanding builds on the theoretical foundation of concepts of resilience outlined earlier in Chapter 2 and the dynamics of protracted displacement and fragility introduced in Chapter 3, which follows the principles of the purposeful transformation of social-ecological systems and takes into consideration the engagement of city officials and key stakeholders as part of the resilience "system" which is investigated further in Chapter 7, to influence the global environmental drivers of climate change at the local level and better define the role of key stakeholders in adaptation to and mitigation of climate change towards building urban resilience (Harris *et al.*, 2014; Sellberg *et al.*, 2015, Quinlan *et al.*, 2016).

However, measuring resilience is more inter-linked with the quantitative methods and numeric values used to measure the capacity of city systems to accommodate the rights and needs of IDPs and refugees in fragile settings. In this case, identifying the indicators of the measuring instrument can lead to creating stronger inter-linkages with the 2015–2030 global agendas (Quinlan & Peterson, 2012) with reflections on using measurements of resilience in development as a means for building capacity among institutional bodies and local communities, as applied in this chapter by using the UNDRR Disaster Resilience Scorecard. In understanding the inter-linkages between measuring and assessing resilience, the Making Cities Resilient Campaign and the Scorecards Ten Essentials are explored to investigate the level of representation of DRR key stakeholders in the process of assessing and measuring resilience.

In 2010, the Making Cities Resilient Campaign was developed by UNDRR and its partners to assist local governments in assessing their progress in building resilience to disaster. It is part of a series of methods for measuring the progress of nations and communities towards meeting the objectives of the Hyogo Framework for Action 2005–2015 (HFA) to increase understanding and encourage commitment by local and national governments to make disaster risk reduction and resilience a priority of policy and to bring the global Hyogo Framework closer to local need. This was followed by the launch of the Local Government Self-Assessment Tool (LG-SAT) in April 2012, in support of the global Making Cities Resilient Campaign, to enrich understanding of disaster risk and to identify gaps in planning policies and financial risk investments.

Being the first method for assessing resilience launched in April 2012, the Local Government Self-Assessment Tool (LG-SAT) was selected as the focus of this book, considering its prominent role in support of the implementation of the 2005–2015 Hyogo Framework for Action (HFA). At the local level, more than 1,850 cities from 95 countries participated in the 2010 Making Cities Resilient (MCR) Campaign. This is an online method in which the main "Ten Essentials" are identified, which is a set of indicators that serve as a baseline measurement that helps cities to assess their current level of disaster resilience. Broad in scope, 293 cities from 13 countries in the MENA Region (Arab States) signed.

Of 40 cities identified as role models to share knowledge from which to learn, only four were in the MENA Region, including Dubai (United Arab Emirates), Aqaba (Jordan), Beirut, and Byblos (Lebanon). These figures indicate the necessity to identify the gaps in processing the LG-SAT tool in the

Arab States of the MENA Region, which is investigated further in this chapter, to understand the methodological framework for progress in assessing resilience, monitoring resilience action plans and feedback mechanisms for local governments (UNDRR, 2013).

With this in mind, New Ten Essentials were launched at the 2017 Global Platform for Disaster Risk Reduction in Cancun, Mexico. Aimed at assisting countries and local governments in monitoring and reviewing progress and challenges in the implementation of the SFDRR, the New Ten Essentials provide two levels of preliminary and detailed assessments. While responding to key questions in a multi-stakeholder exercise is suggested on both levels, the detailed assessment includes 117 quantitative indicators that can serve as the basis for a detailed action plan for city resilience. Considering that the indicators of the scorecard were developed based on a "scientifically rigorous", top-down methodology, it is difficult to justify the inclusion of citizen-based knowledge and community participation.

In support of the argument above, it is noted that, according to a global online survey by UNDRR in 2018, the UNDRR Local Government Self-Assessment Tool (LG-SAT) is the most frequently used tool by local governments, followed by the Preliminary and Detailed Disaster Resilience Scorecards for Cities. There were 159 valid responses to the survey which entailed 58 questions about the local governments, local risks, and understanding of risk, risk communication, local DRR strategy, strategy implementation, and DRR actions and experience (UNDRR, 2019). Figure 4.1 shows the impact of using the MCR campaign as an awareness-raising mechanism to increase the outreach of assessing resilience by the top-four cities. The use of the City Resilience Profiling and City Resilience Index tools in the Arab States of the MENA Region is explored further in the study to accumulate evidence about the long-term impact of such tools in fragile settings, as "just because capacities, knowledge or networks exist does not mean that they will be accessible or useful in a given crisis situation" (Schipper & Langston, 2015).

The process of assessing resilience using the Disaster Resilience Scorecard starts with a self-assessment process led by city leaders before sharing results with the UNDRR office. To support the knowledge-sharing process between cities, the MRC Campaign was introduced as a platform to facilitate the learning exchange between cities and sharing experiences on strengthening risk governance and DRR institutional strategies through self-assessments of resilience.

In order to ensure the connectivity between the local municipalities joining the campaign and the legitimacy of the self-assessment process, the UNDRR encourages the offices of city mayors to seek the approval of the city council to be involved officially as a "participating city" in the campaign. The local government informs the central government about the participation and notifies the focal point of the official Hyogo Framework for Action or the National Platform for Disaster Risk Reduction. Accordingly, the city is added to the

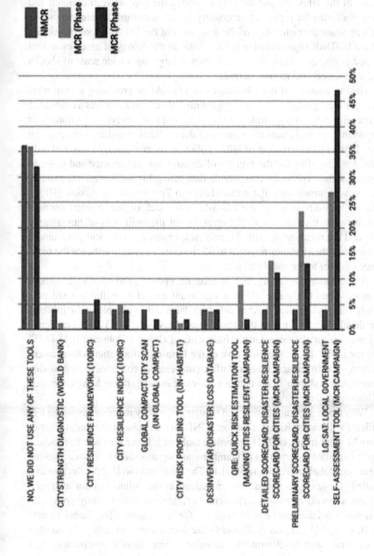

Figure 4.1 Different tools used by the local governments to support DRR strategies (UNDRR, 2019)

web-based campaign city map and the local government can create its own online profile on the campaign website. On a scale of 1–5, the Local Government Self-Assessment Tool includes Ten Essentials and 41 local-context indicators. The key questions are aligned to the priority areas and core indicators of the HFA. Paving the way to bridge the gap between national and local platforms for reporting on country progress towards resilience, one of the important characteristics of the indicators of the LG-SAT tool mentioned by the UNDRR representative is its holistic methodology of assessment that helped to engage a large number of cities in targeting a wide scale of shocks and stresses beyond natural hazards.

Having considered the advantages of LG-SAT in providing a joint platform for local governments to engage with different stakeholders as indicated in the qualitative and quantitative data analysed in the previous sections, it is important to complement the process of data collection and map the existing gaps in the implementation of DRM policy at the city level, as this tool only provides a baseline for the process of disaster risk governance and helps to develop status reports for comparable data across local governments in association with monitoring the national Hyogo Framework for Action (HFA). The main disadvantage of the LG-SAT is the lack of fundamental connection between the mathematical expression of the indicators of quantitative data and the relationship with the empirical observation of local government context and the solutions required to fill the existing gaps to enhance the resilience of cities to the risk of disaster.

At the national level, the lack of accurate, updated, and reliable data about disaster losses in the region has a significant impact on building resilience. "Only nine out of the 22 Arab countries have either completed or initiated the development of national disaster loss" (UNDRR, 2013). This evidence is based on the database for the International Disaster Database (EM-DAT) and the multi-stakeholder initiative of the Disaster Information Management System – DesInventar. This is a platform that enables countries to analyse disaster trends and their impacts in a systematic manner, through the collection of historical disaster data.

With the lack of monitoring and regular updates by member states, the UN Office for Disaster Risk Reduction (UNDRR) launched the Sendai Framework Monitor in March 2018, an online tool that is based on a set of 38 indicators, to monitor the progress of member-state signatories to towards achieving the seven global targets of the SFDRR. The Disaster Loss Data Collection tool (called "DesInventar Sendai") is embedded in the online monitor as a subsystem, allowing national governments to create and maintain fully-compliant loss databases that can be used to gather the data required for Global Targets A, B, C, and D. Targets E, F, and G are focused on increasing the number of countries with DRR strategies to enhance international co-operation and increase the availability of and access to multi-hazard early warning systems. Figure 4.1 provides an overview of the seven-step approach required for national governments to support the monitoring process of disaster data loss at the national level (Clarke *et al.*, 2018).

Despite these efforts, the 2021 Arab States Regional Assessment Report (RAR) indicated that in reviewing progress, SFM, DesInventar1 and Emergency Events Database (EM-DAT) resources are used as only 10 countries have provided national disaster loss data, albeit intermittently. Aligned with the four priorities of the SFDRR for Action (articulated in the central diagram) – Priority 1: Understanding disaster risk; Priority 2: Strengthening disaster risk governance to manage disaster risk; Priority 3: Investing in disaster risk reduction for resilience; and Priority 4: Enhancing disaster preparedness for effective response and to "Build Back Better" – are four targets outlined on the horizontal level: (a) Substantially reduce global disaster mortality by 2030, aiming to lower the average per 100,000 global mortality rate in the decade 2020–2030 compared with the period 2005–2015; (b) Substantially reduce the number of affected people globally by 2030, aiming to lower the average global figure per 100,000 in the decade 2020–2030 compared with the period 2005–2015; (c) Reduce direct economic loss caused by disaster in relation to Global Gross Domestic Product (GDP) by 2030; and (d) Substantially reduce disaster damage to critical infrastructure and disruption of basic services, among them health and educational facilities, through developing their resilience by 2030.

Building Urban Resilience

This book provides an attempt to frame a conceptual approach to integrating adaptation to climate change into the formulation of the Urban Resilience Action Plan (U-RAP), generating inter-linkages between the categorised variables of the U-RAP and the Ten Essentials to build and operationalise urban resilience. The urgency of this study comes from the time frame assigned to achieve COP27 targets and to increase substantially the number of countries updating their National Adaptation Programmes of Action (NAPAs), while integrating them into their strategies to reduce national and local risk of disaster before COP28. This cannot be achieved without building coherence between the 2015–2030 Global Framework at the local level, identifying the mechanisms for shared data collection, and combining institutional capacities for developing Resilience Action Plans. The literature about resilience is reviewed in the contexts of climate security displaced people outlined earlier in Chapter 3, with the aim of assessing the impact of new UNDRR national and local tools and monitoring systems in building urban resilience and bridging the gap between national and local DRR platforms. The inter-linkages between DRR operational scales are explored, keeping in mind that the Sendai Framework Monitor is related to global targets of the SDGs, particularly SDG 11 (sustainable cities and communities). This will help to investigate the inter-linkages with measures of mitigation and adaptation to climate change and further envisage the theoretical underpinnings behind the framework.

Known as the process of assembling different components into one structure, building resilience is the progress from measuring resilience to assessing resilience and the actions that follow (Holling, 1978). From a socio-ecological perspective of resilience, seven strategies have been outlined by Quinlan *et al.* (2016) to build resilience: maintain diversity and redundancy; manage connectivity; manage slow variables and feedbacks; foster an understanding of social-ecological systems as complex adaptive systems; encourage learning and experimentation; broaden participation; and promote poly-centric governance systems (Biggs *et al.*, 2012).

These strategies for enhancing resilience can be structured to focus primarily on the resilience of the social-ecological system or its governance and by whether the focus on resilience is based on the structure of the system or its dynamics. These strategies are complementary and can be combined, emphasising the level of engagement of DRR key stakeholders in the process of assessing resilience, which can affect the efficacy of the process management and validity of the assessment results, both required for enhancing the development of Urban Resilience Action Plans.

Measures of Climate Change Mitigation and Adaptation

In understanding the inter-relationship between these two concepts, and how they are associated with urban resilience in the context of the MENA Region, it is important to investigate the systemic nature of risk and how it evolves in assessing the wider set of cascading impacts of climate change and underlying inter-dependencies the drivers of risk, together with the spatial extent of systemic risk through assessments of compound climate risk. The UNDRR indicates that "assessing interconnectedness and non-linearity in the cause – effect relationship of systemic risk and identifying critical sub-systems within the system that have the potential to fail or even reach a tipping point, are essential attributes in evaluating systemic risk" (Schweizer & Renn, 2019, cited in UNDRR, 2022).

Mitigating climate change means reducing the flow of heat-trapping greenhouse gases (GHGs) into the atmosphere. Since mitigation is associated with reducing the emissions or enhancing the sinks of greenhouse gases that are generated from all human activities, such as carbon dioxide, methane, and nitrous oxide, the spatial and temporal scale of the effects of mitigation cannot be achieved without a transitional shift towards a low-carbon economy. Despite the pledges of countries to reduce their emissions, in the annual UNEP Emissions Gap Report 2022, it was indicated that "global annual GHG emissions must be reduced by 45 per cent compared with emissions projections under policies currently in place". To be on track to limit global warming to 1.5°C, this target must be achieved in just eight years. The sustainability of mitigation actions such as changes in energy sources, industrial processes and farming methods, and the formulation of new policies such as carbon pricing by carbon taxes and carbon emission trading, must continue rapidly after 2030 to avoid exhausting the limited remaining atmospheric carbon budget (UNEP, 2022).

In measuring climate mitigation, several methods have been generated to help to assess the economic impact of mitigation measures and drive the global transition towards a low-carbon economy such as the Climate Change Indicators Dashboard, a statistical tool developed by the International Monetary Fund to help to link climate considerations and global economic indicators. Providing information at country level about the contribution of renewable energy in nationally determined contributions (NDCs) of countries, efforts to localise data at city level have also taken place using other tools, such as the Financial Stability Board, which created the Task Force on Climate-related Financial Disclosures (TCFD) to improve and increase reporting of climate-related financial information.

Aimed at "understanding the risks that climate changes pose to city finances, integrating them into city plans and disclosing them publicly can help to build support for climate action, boost funding for low-carbon projects, and mainstream climate into the city's financial processes", defining the impact on building resilience for local communities in fragile contexts remains missing in evidence for how donor-led investments in humanitarian aid can help to shift "repetitive and reactive humanitarian response, towards a forward-looking and proactive risk management model" (WFP, 2022). Anticipatory and early actions regarding climate hazards helped to support communities in the rural context and helped to reduce human and capital losses caused by severe climate-related weather and extreme events; monitoring impact in the urban setting is explored further in Chapter 5.

Strongly associated with financial risks, cities disclose and report on their progress towards mitigation of climate change by monitoring their governance, strategy, risk management, metrics, and targets, yet, for climate adaptation, policies remain difficult to monitor because of the complexities associated with the valuation and monetisation of costs, benefits, and returns of the long-term impact of climate adaptation (Bours *et al.*, 2013). "Monitoring the performance and scalability of climate adaptation finance remains difficult due to challenges associated with context dependency, confidentiality restrictions, uncertain causality, and a lack of unified and agreed-upon impact metrics" (Richmond & Hallmeyer, 2019).

Tackling these challenges, the Office of Strategic Planning and Development Effectiveness at the Inter-American Development Bank developed Impact-Evaluation Guidelines aimed at developing reasonable indicators of adaptation and mitigation for climate change projects in association with the wider sustainable development agenda. Defining a four-stage approach of data collection, data analysis, reviewing the existing legal and regulatory frameworks, and project piloting, which can help in the formulation of a National Adaptation Programme of Action (NAPA), and developing appropriate, specific indicators for outputs and outcomes, which can help to capture co-benefits that might only be realised after the life of the project, as well as having intermediate indicators that are clearly linked to near-term benefits, the guidelines make attribution for assessing impacts easier in future (McCarthy *et al.*, 2012).

Reference list

Bahadur, A., Wilkinson, E., & Tanner, T. M. 2015. Measuring resilience: An analytical review. *Climate and Development*.

Berkes, F. 2000. *Linking social and ecological systems*. Cambridge University Press.

Biggs, R., Schlüter, M., & Schoon, M. L. (Eds.). 2015. *Principles for Building Resilience: Sustaining Ecosystem Services in Social-Ecological Systems*. Cambridge University Press.

Bosetti, L., Ivanovic, A., & Munshey, M. 2016. Fragility, risk, and resilience: A review of existing frameworks. *Background Paper: 1–12*. UN University Centre for Policy Research.

Bours, D., McGinn, C., & Pringle, P. 2013. *Monitoring & Evaluation for Climate Change Adaptation: A Synthesis of Tools, Frameworks and Approaches*. SEA Change Community of Practice and UKCIP.

Clarke, L., Blanchard, K., Maini, R., Radu, A., Eltinay, N., Zaidi, Z., & Murray, V. 2018. Knowing what we know – Reflections on the development of technical guidance for loss data for the Sendai framework for disaster risk reduction. *PLoS Currents*, 10.

Davoudi, S., Shaw, K., Haider, L. J., Quinlan, A. E., Peterson, G. D., Wilkinson, C., Fünfgeld, H., McEvoy, D., Porter, L., & Davoudi, S. 2012. Resilience: a bridging concept or a dead end? "Reframing" resilience: challenges for planning theory and practice interacting traps: resilience assessment of a pasture management system in Northern Afghanistan urban resilience: what does it mean in planning practice? Resilience as a useful concept for climate change adaptation? The politics of resilience for planning: a cautionary note: edited by Simin Davoudi and Libby Porter. *Planning theory & practice*, 13(2): 299–333.

Expert from the Overseas Development Institute. 2018. (Interview, 16 April).

Financial Stability Board. 2023. *Task Force on Climate-related Financial Disclosures*. Available at: www.fsb-tcfd.org/

Food Security Information Network (FSIN). 2016. *Measuring Resilience*. Available at: www.foodsecurityportal.org/measuring-resilience

Holling, C. S. 1973. Resilience and stability of ecological systems. *Annual Review of Ecology and Systematics*, 4(1): 1–23.

International Monetary Fund. 2023. *Climate Change Indicators Dashboard*. Available at: https://climatedata.imf.org/

Kerner, D. A., & Thomas, J. S. 2013. Resilience attributes of social-ecological systems: Framing metrics for management. *Resources*, 3(4): 672–702.

McCarthy, N., Winters, P., Linares, A. M., & Essam, T. 2012. *Indicators to Assess the Effectiveness of Climate Change Projects*. Environmental Science.

Quinlan, A. E., Berbés-Blázquez, M., Haider, L. J., & Peterson, G. D. 2016. Measuring and assessing resilience: Broadening understanding through multiple disciplinary perspectives. *Journal of Applied Ecology*, 53(3): 677–687.

Richmond, M., & Hallmeyer, K. 2019. *Tracking Adaptation Finance: Advancing Methods to Capture Finance Flows in the Landscape*. A CPI Climate Finance Tracking Brief. Initiative, CP.

Schipper, L., & Langston, L. 2015. *A Comparative Overview of Resilience Measurement Frameworks – Analysing Indicators and Approaches – Working and Discussion Papers*. Overseas Development Institute.

Schweizer, P. J., & Renn, O. 2019. Governance of systemic risks for disaster prevention and mitigation. *Disaster prevention and management: an international journal*, 28(6): 862–874.

Sellberg, M. M., Wilkinson, C., & Peterson, G. D. 2015. Resilience assessment: a useful approach to navigate urban sustainability challenges. *Ecology and Society*, 20(1).

Sharifi, A. 2016. A critical review of selected tools for assessing community resilience. *Ecological Indicators*, 69: 629–647.

Thomas, S., & Kerner, D. 2013. Metrics for assessing resilience: A social-ecological systems perspective. In: *Proceedings of the Presentation to the Challenges of Natural Resource Economics and Policy National Forum on Socioeconomic Research in Coastal Systems*, p. 25. New Orleans, LA. Stockholm Environment Institute and Stockholm Resilience Centre.

UNDRR. 2013. *Factsheet: Overview of Disaster Risk Reduction in the Arab Region.* United Nations Office for Disaster Risk Reduction. Available at: www.unisdr.org/we/inform/publications/31693

UNDRR. 2019. *Making Cities Resilient (MCR) Campaign: Comparing MCR and Non-MCR Cities.* United Nations Office for Disaster Risk Reduction. Available at: www.unisdr.org/campaign/resilientcities/toolkit/article/report-on-the-making-cities-resilient-mcr-campaign-comparing-mcr-and-non-mcr-cities

UNDRR. 2022. *Technical Guidance on Comprehensive Risk Assessment and Planning in the Context of Climate Change.* United Nations Office for Disaster Risk Reduction.

UNEP. 2022. *Emissions Gap Report 2022.* United Nations Environment Program. Available at: www.unep.org/resources/emissions-gap-report-2022

Walker, B., Carpenter, S., Anderies, J., Abel, N., Cumming, G., Janssen, M., Lebel, L., Norberg, J., Peterson, G. D., & Pritchard, R. 2002. Resilience management in social-ecological systems: A working hypothesis for a participatory approach. *Conservation Ecology*, 6(1).

Walker, B., & Salt, D. 2012. *Resilience Practice: Building Capacity to Absorb Disturbance and Maintain Function.* Island Press.

WFP. 2022. *Climate Risk Financing. Anticipatory and Early Actions for Climate Hazards.* World Food Program.

5 Monitoring Urban Resilience

Data: Multi Hazard Early Warning Systems (MHEWS) and Climate Information

It might seem plausible to argue that national monitoring of disaster data losses can help to achieve progress in reporting to the Sendai Framework for Disaster Risk Reduction (SFDRR), the Climate Change (CC) Agenda, the Sustainable Development Goals (SDGs), and the Habitat III Agenda global targets. Nevertheless, with the lack of climate security protracted displacement data in the MENA Region, the complexity of urban disaster, urban conflict, and urban poverty, exacerbates the exclusion of internally displaced persons (IDPs) and refugees from assessments of resilience to disaster, which remains an obstacle in achieving the 2015–2030 global targets at the local level.

According to the World Meteorological Organisation Report 2021, the number of disasters has increased by a factor of five over a 50-year period, shown in Figure 5.1 (WMO, 2021). In the past decade alone, approximately 700,000 people have lost their lives, over 1.4 million have been injured, and 23 million have been made homeless because of disasters. At the same time, the "Hyogo Framework for Action (HFA) 2005–2015: building the resilience of nations and communities to disasters" was adopted by the World Conference on Disaster Reduction, but the layer of extensive risks was "not captured by global risk modelling, nor are the losses reported internationally" (UNDRR, 2015). Developments in this field took place with the launch of the UNDRR Comprehensive Disaster and Climate Risk Management (CRM) Programme, noting that climate adaptation does not prevent all losses and damages and that capturing "information from weather, seasonal and climate forecasts and predictions, and translating such information into meaningful information" is critical to build resilience to risks across different timescales – short, medium, and long-term – to help to achieve more comprehensive planning and implementation of national and sub-national strategies for disaster risk reduction and integrate risk-centred approaches into National Adaptation Plans (NAPs) (UNDRR, 2022).

DOI: 10.4324/9781003363224-5

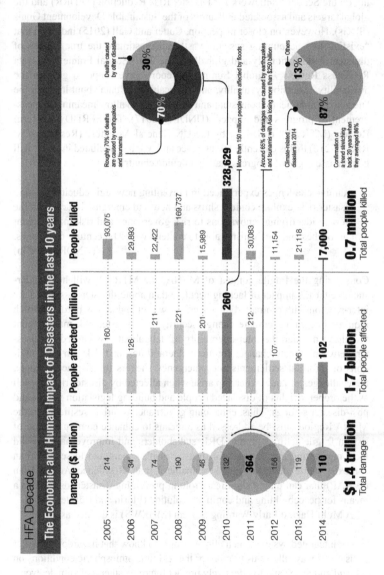

Figure 5.1 The Economic and Human Impact of Disasters in the last HFA Decade (UNDRR, 2014)

At first sight, it might seem plausible to argue that effective monitoring of data about loss caused by disaster can help to achieve progress in reporting on the Sendai Framework for Disaster Risk Reduction (SFDRR) and the global targets and associated indicators of the Sustainable Development Goals (SDGs). However, on closer inspection, Cutter and Gall (2015) indicated that "existing loss accounting systems vastly underestimate the true burden of disasters, both nationally and globally". The 2017 Sendai Framework Data Readiness Review – Global Summary Report, gave scope to gaps in the availability, accessibility, quality, and applicability of data about loss and the "need to be sufficiently consistent and comparable to allow meaningful measurement of progress and impact" (UNDRR, 2017). The SFRDD Mid-Term Review (MTR SF), called for by the UN General Assembly (Resolution A/RES/75/216) that is currently taking place and will be concluded in 2023, will help to assess global progress made and guide countries to:

> examine challenges experienced in preventing new and reducing existing disaster risk, explore context shifts and new and emerging issues, with the view to identifying renovations to risk governance and risk management approaches and mechanisms able to contend with 21st century challenges.
>
> (UNDRR, 2023a)

Considering the fragile context of MENA, the MTR SF will help understand better the impact of lacking accurate data about disaster losses and the inter-relation with protracted displacement within and across borders, which requires integrating climate change security into disaster risk management to help to manage the inter-operability of data about loss caused by disaster, between climate change, conflict vulnerabilities, and risks. In the conditions of informal settlements and emergency shelters for displaced people, the challenges of land ownership arise when affected by disaster risk, as well as the property rights of displaced people and building legislation in pre- and post-disaster settings. Thus, enhancing governance of urban resilience in the MENA Region must be addressed as a means to enhance communication of risk in fragile settings among DRR stakeholders and improve national and local monitoring mechanisms for disaster displacement and losses in urban contexts, while developing a sustainable Urban Resilience Action Plan (U-RAP) that can play a significant role in preventing disaster loss and as a means for peacebuilding and conflict resolution (Florin and Bürkler, 2017).

A Multi Hazard Early Warning System (MHEWS) is defined as:

> An integrated system which allows people to know that hazardous weather is on its way, allows us to monitor the real time atmospheric conditions on land and at sea and to effectively predict future weather and climate events using advanced computer numerical models, and informs how governments, communities and individuals can act to minimise the impending impacts.
>
> (UNFCCC, 2022)

Composed of four main components: risk knowledge, observation, communication, and response, as shown in Figure 5.2, an efficient and sustainable MHEWS is people-centred to empower the preparedness and anticipatory decision-making process of the most vulnerable. Developing legislation and regulatory frameworks for emergency response associated with the complexity of hazards in fragile settings in MENA is critical to ensure that inclusive risk communication practices are applied to minimise anticipated impacts and reduce vulnerabilities to climate.

The role of Early Warning Systems in reducing the impact of disasters is noted. In the 2022 UNDRR-WMO joint report on the global status of multi-hazard early warning systems (MHEWS) against Target G of the Sendai Framework, it is indicated that "cascading human and economic cost of high-frequency, low-impact events continue to rise, with the need to advance and accelerate MHEWS at all levels" (UNDRR & WMO, 2022).

In March 2022 at the World Meteorological Day, the UN Secretary-General, António Guterres, set a new target to be aligned with SFDRR, indicating that "within the next five years, everyone on Earth should be protected by early warning systems against increasingly extreme weather and climate change" (UNFCCC, 2022). This task was led by the World Meteorological Organisation (WMO) to present an action plan at the UN COP27 climate conference in Egypt.

Noting that one third of the world's people, mainly in least-developed countries and developing small island states, are still not covered by early warning systems (60% of the population lack coverage in Africa alone). Efforts to bridge the gap between national and local governments in developing anticipatory response plans, building capacities, and directing investments towards comprehensive early warning systems are lagging behind. This is strongly evident in Figure 5.3, where the results from the Sixth Workshop of the Glasgow – Sharm el-Sheikh Programme about the global goal for Climate Change Adaptation (CCA) "Target-setting, metrics, methodologies and indicators", highlighted the gap in reporting on having MHEWS in the MENA Region, which reflects the need for urgent action at local and national levels.

The Monitoring and Early-Warning System for climate change was recognised as one of the key pathways of Climate Adaptation for sustainable ecosystems in the IPCC 6th Assessment Report (Figure 5.4). Associated with its impact on increasing the resilience of bio-diversity and ecosystem services to climate change, MHEWS can help to minimise negative impacts, protect, and inform anticipatory management intervention. Inter-linkages with socio-economic resilience are also indicated by the role of MHEWS in directing investments towards social safety nets, as well as reducing loss and damage by spreading and sharing risk that can help to advance mechanisms for financing CCA. MHEWS also plays a major role in informing science-based and standardised disclosure of data about climate, to guide assessments of climate risk, and to create the enabling conditions for interventions in making the transition to green mitigation of climate change.

Detection, observations, monitoring, analysis and forecasting of hazards

Develop hazard monitoring and early warning services

- Are the right parameters being monitored?
- Is there a sound scientific basis for making forecasts?
- Can accurate and timely warnings be generated?

Warning dissemination and communication

Communicate risk information and early warnings

- Do warnings reach all of those at risk?
- Are the risks and warnings understood?
- Is the warning information clear and usable?

Disaster risk knowledge

Systematically collect data and undertake risk assessments

- Are the hazards and the vulnerabilities well known by the communities?
- What are the patterns and trends in these factors?
- Are risk maps and data widely available?

Preparedness and response capabilities

Build national and community response capabilities

- Are response plans up to date and tested?
- Are local capacities and knowledge made use of?
- Are people prepared and ready to react to warnings?

Figure 5.2 Four components of an early warning system (WMO, 2022)

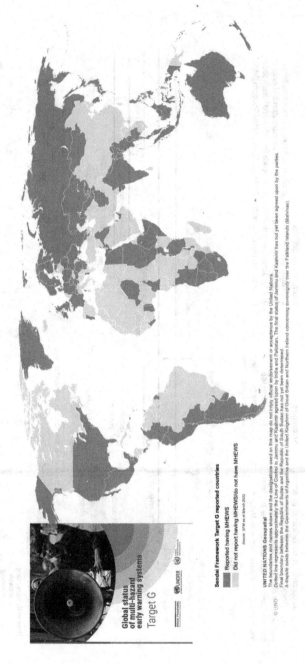

Figure 5.3 Early Warning Systems – Target G. (UNDRR, 2023)

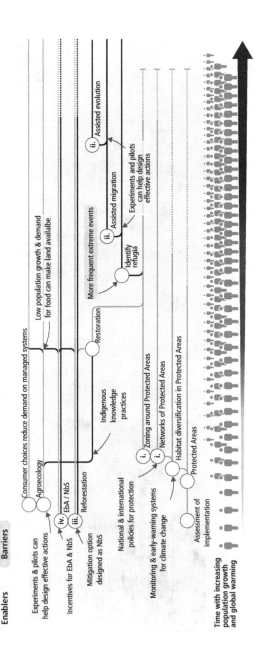

(d) Adaptation pathways for ecosystems.

Adaptation options can be facilitated by actions which increase the solution space such as consideration of local knowledge, new regulations and incentives but also decrease due to climatic and non-climatic stressors and maladaptation

Strategies
— Protect
— Restore/migrate
— Sustainable use
···· Uncertainty in effectiveness with increasing pressures

Examples for actions
i. Networks of Protected Areas combined with zoning increase resilience.
ii. Assisted migration and evolution might reduce extirpation and extinction.
iii. Adaptation and mitigation increase space for nature and benefit society.
iv. Ecosystem-based Adaptation (EbA) and Nature-based Solutions (NbS).

Enablers   Barriers

Experiments & pilots can help design effective actions

Incentives for EbA & NbS

iv.   EbA / NbS
iii.   Reforestation

Mitigation option designed as NbS

National & international policies for protection

Monitoring & early-warning systems for climate change

Assessment of implementation

Time with increasing population growth and global warming

Consumer choices reduce demand on managed systems
Agroecology

Low population growth & demand for food can make land availalbe

Indigenous knowledge practices

Restoration

i.   Zoning around Protected Areas
i.   Networks of Protected Areas

Habitat diversification in Protected Areas

Protected Areas

Identify refugia

More frequent extreme events

ii.   Assisted migration

Experiments and pilots can help design effective actions

ii.   Assisted evolution

Figure 5.4 Adaptation pathways for ecosystems (IPCC, 2022)

Accordingly, integrating climate change adaptation in generating U-RAPs from the indicators of the New Ten Essentials, introduced in Chapter 4, will help to understand using Open Data for monitoring progress on data about losses caused by disaster and conflict and using the SFDRR as a means for conflict resolution. Comparative analysis of tools and indices for resilience to disaster also helped most to identify the gaps and opportunities for integrating principles of societal resilience of IDPs and refugees into DRR policy of prevention and protection, adopting the recommendations of the 2016 High-level Independent Panel on Peace Operations (HIPPO) for the UN system to "pull together in a more integrated manner in the service of conflict prevention and peace" (Stein & Walch, 2017).

In the conflictual and most fragile settings of the MENA Region, developing effective, community-led methods of communicating warnings is critical with the absence of institutional capacities and disruption of formal mechanisms for disaster risk reduction. When aligning disaster risk communication with local governance structure, it is important to fit with multi-lingual environments and to ensure that the process of decision-making about urban resilience takes place through trusted channels of communication and authoritative sources to maintain inclusivity of all vulnerable groups (UNEP, 2015).

Urban Resilience Governance: Disaster Risk Reduction Policies and Local Capacities

Governance, as defined by the United Nations Development Programme (UNDP), is "the exercise of political, economic and administrative authority in the management of a country's affairs at all levels". Urban resilience comprises a system where mechanisms, processes and institutions operate to define actions and identify finances required to build physical and socio-economic resilience. By articulating the needs and interests of citizens and most disadvantaged groups, governance of urban resilience helps them to exercise their legal rights, meet their obligations, and mediate their differences. "Governance encompasses, but also transcends, government. It encompasses all relevant groups, including the private sector and civil society organisations" (Cheema, 2020).

In addressing human security in the context of justice regarding CCA the inclusion of human rights laws is critical to maximise the benefits of risk governance at local level, with the legislation and regulation of policies formed concerning climate change, DRR, and non-binding global frameworks for sustainability. In the conditions of informal settlements and emergency shelters for displaced people, the challenges of land ownership arise, thus the property rights of displaced people and building legislation in pre- and post-disaster settings, urban politics, and urban governance are emphasised in this chapter to support understanding of the methodological approaches to measuring resilience using the Sendai Framework for Disaster risk reduction (SFDRR), capturing social and bio-physical vulnerabilities and building coherence with the 2015–2030 SDGs.

The understanding of risk in fragile settings is crucial to investigating the principles of building the governance of urban resilience, to help to define the actions required, and to prioritise investments in climate change mitigation and adaptation for those who are most in need. The International Risk Governance Council defines "risk" as an uncertain consequence of an event or activity in relation to something endowed with human value (Florin and Bürkler, 2017). Supported by the perception of risk as "the probability of a loss, and depends on three elements, hazard, vulnerability, and exposure" (Crichton, 1999), understanding the dependency between these three elements in increasing or decreasing risk can be associated with the climate prediction models and EWS to guide the governance of urban resilience.

"DRR is a complex cross-cutting issue that requires an interdisciplinary and multi-level approach, bringing together knowledge, skills and resources of different stakeholders" (UNDP, 2020). Thus, data play a major role in determining an approach to urban resilience to multi-hazard and processes for inclusive risk-informed decision-making, reducing vulnerabilities and exposure to risk while "integrating resilience into socio-economic development planning and infrastructure will safeguard development investments" to achieve SDG Goal (11) (UNDRR, 2017).

With a focus on the urban risk profile, history of conflict, and highlighting the underlying drivers of risk, urban poverty, weak urban governance, and political stability in the most vulnerable states in the MENA Region, it is critical to introduce the concept of fragility in terms of conflict at the local level to address how National Adaptation Plans and strategies for disaster risk reduction are disconnected from local mechanisms for adaptation to climate change and the need to embed socio-ecological principles and political drivers of community-led urban resilience in the process of DRR decision-making at the local level.

In humanitarian settings after disaster and conflict, data are collected from different sectors, NGOs, and international donor organisations, depending on their scope of work and area of action. Nevertheless, with the lack of monitoring of data about losses caused by disaster and verification of information about exposure to vulnerability by governmental bodies and responsible local institutes hinder the directing of urban resilience efforts and financing of humanitarian aid towards those who are most affected. Understanding the socio-political context of governance and the structure of public finance will help to overcome divisions and conflict in developing and implementing strategies and early warning disaster prevention systems to mitigate the risk of disaster (Cheema, 2022).

Decision-Making: Urban Resilience Action Planning (U-RAP) and Policy Formulation

An action plan is a set of actions that describes the way in which an organisation can meet its objectives through detailed action steps that describe how and when these steps will be taken. Developing an action plan can help decision-makers about change to turn visions into reality and increase efficiency and accountability within an organisation. Reflected in the process for decision-making about resilience, this philosophical thinking was framed following the investigation of tools and frameworks for assessing resilience, applied in Chapter 4, to define the steps of reporting on measures of resilience (indicators) and comprehending how and when to help to shift the organisational system of a city from assessment to implementation.

There have been several attempts to develop scientific-based indicators to enable monitoring of the processes of resilience-building and to establish an evidence-based mechanism to help to "make resilience more tangible for decision and policy makers as well as society at large" (Feldmeyer *et al.*, 2019). Nevertheless, without having the legislative systems, legal frameworks, and financing mechanisms that could enable direct investments in implementing resilience-building, it is not possible to monitor the impact of short- and long-term adaptation to climate change to govern and steer processes for sustainability transformation in the MENA Region (Kumar, 2022).

Gall (2013) stressed the complexities associated with this process because "putting a framework into action requires the selection of indicators, identification of feedback loops and so forth. Frameworks are a great starting point but many decisions on how to implement the model and measure resilience are left unresolved". Owing to the lack of mechanisms for stakeholder engagement and lack of understanding of the local context of the end-user, the process of decision-making about resilience was investigated in the study using the SFDRR New Ten Essentials as the variables affecting the U-RAP process of decision-making.

Figure 5.5 shows a diagram developed by the UNDRR that explains the ideal process for developing an Urban Resilience Action Plan (U-RAP). Divided into five main vertical sections, the process of decision-making about urban resilience starts from the identification of the characteristics of the system using the New Ten Essentials broken down to a certain number of indicators and sub-indicators for each essential. Sections two and three determine the resilience score, which feeds into identifying the gaps that need to be addressed according to the total result and guides the decision-making process of identifying specific actions to address the gaps and develop the resilience action plan. Nevertheless, evidence from primary data collected for this study indicates the need to understand the complexities behind this process.

The complexity of urban informality in large capital cities in fragile states in MENA is evident in the forced relocations and evictions that persist in the ad-hoc response of the states to the growing demand by IDPs and refugees for access to housing and public services, highlighting the intersection of financing for urban resilience with conflict and displacement caused by disaster. It is proclaimed in several studies that the governments' approach of demolishing and relocating informal settlements that frequently appear in humanitarian settings violates human rights (Burger and Wentz, 2015), yet this must be associated with the lack of

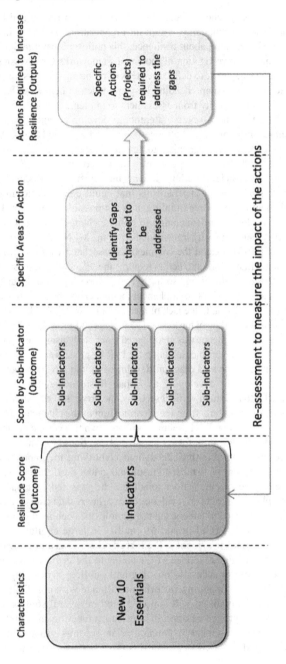

Figure 5.5 Steps for the development of Local DRR Plans (UNDRR, 2015)

pre-financing for disaster prevention and the failure when planning solutions for urban resilience and adaptation to climate to secure access to land supply, rights of residence, security of tenure, and affordable housing required in recovery and reconstruction after disaster and conflict. The relocation of displaced populations to temporary shelters and camps in city outskirts worsens inequalities caused by climate change by trapping IDPs in vicious circles of more squatting in new areas because of prolonged demand for better livelihoods, social networks, and employment opportunities. The lack of finance mechanisms and housing funds obstructs IDPs.

The legislation of human rights for people displaced by climate and disaster is trapped in the contradiction between the "hard laws" for human rights and the "soft laws" for displacement caused by disaster. In a situation of forced displacement caused by protracted conflict and disaster, the absence of rules and principles for directing humanitarian investments towards long-term recovery from climate change can lead to limiting funds for emergency response and ignoring the long-term demand for land tenure and property rights for displaced people. The protection efforts of United Nations agencies mostly provide for people displaced by conflict according to the norms of supporting the settlement of refugees and voluntary return of IDPs, yet, at the city planning level, social integration and access to infrastructural services are only outlined from the broader perspective of humanitarian law for protection, overlooking the long-term impact of enhancing adaptation to climate and resilience-building in fragile settings.

Hence, the key question in this chapter is: How can the complementary integration of "hard laws" for human rights support the financial investment in urban resilience in fragile settings and the enforcement of "soft laws" for disaster risk reduction protect land and property rights of people displaced by disaster in the MENA Region? With the understanding that "hard-law instruments allow states to commit themselves more credibly to international agreements" (Shaffer & Pollack, 2009), the Guiding Principles on Internal Displacement are explored further in Chapter 8 to help to develop operational mechanisms for reporting according to the Sendai Framework for Disaster Risk Reduction (SFDRR), and take actionable measures to strengthen the legally binding nature and obligations of CCA while protecting the rights of IDPs and refugees in pre- and post-disaster and conflict settings.

Noting the role of DRR key stakeholders in urban resilience and the process of decision-making about adaptation to climate, it is important to differentiate between their level of involvement in the two actions of "measuring" and "building" urban resilience, as this will clarify the understanding of the level of responsibilities and roles of DRR key stakeholder involvement. Responsibilities for measuring resilience are associated with local governments and humanitarian aid agencies who have the upper hand in the process of decision-making about resilience, accessibility to the data sets required, and power to approve budget allocation and prioritise DRR investments. On the other hand, this does not waive the responsibility of the IDPs and refugees to understand risk and play an effective role in implementing DRR legislation and preventive measures to build resilience at the grassroots level. Thus, building resilience highlights their role as informal role-players in the building process and the need to involve them better in the decision-making process to strengthen the act of inclusion in the development of an Urban Resilience Action Plan and ownership of the implementation and long-term monitoring of the U-RAP.

The overall subjectivity of the approaches to assessing urban resilience is noted and guides the recommendations for the U-RAP Policy Guidance to adopt an inclusive, community-based, participatory approach to recruiting participants to join the process of assessing the resilience of cities. This will help to improve the validity, inclusivity, and impact of the results being fed into DRR legislative frameworks and action plans for the resilience cities. This bias limits the inclusion of marginalised groups of IDPs and refugees in the process of decision-making about the assessment of and action plan for resilience.

The type of data that is fed into developing and advising U-RAPs and the mechanisms of engaging key actors are also critical components in guiding decision-making processes for adaptation to climate change. Despite the dependence on modelling climate data, quantitative risk assessments and qualitative knowledge-sharing in assessments of resilience to climate, the strategy for sampling design applied in selecting key stakeholders and developing U-RAPs cannot be generalised to the wider population of DRR stakeholders. The identification of participants should take into consideration their professional backgrounds, understanding of the mechanisms applied for their engagement, and the quality of data used to rate the indicators of city resilience to help to capture the observations participants on how each indicator feeds into the proposed resilience action plan. "Outputs" are suitably differentiated as the number of constant-quality actions or activities" that are formed by responding to the Disaster Resilience Scorecard indicators and sub-indicators, while "outcomes" are the final state that is influenced by the level of outputs. Outcomes would then correspond to the purpose for which the actions are identified, and outputs to the actions themselves (Schreyer, 2012). In public policy, "output" is usually referred to as an immediate, tangible yield product, yet this might or might not result in a certain "outcome". Thus, the uncertainty of using the Ten Essentials in forming the U-RAP must be addressed further by investigating the level of agreement among DRR key stakeholders with the quality of datasets (accuracy, validity, reliability, timelessness, accessibility, and relevance) as measures that define the process assessing resilience and affect the efficacy and applicability of U-RAP outcome.

Having structured disaster risk and climate resilience across the different levels of DDR governance is another challenge for the decision-making process for an Urban Resilience Action Plan (U-RAP), which is common with other data sets about disaster in various sectors ranging from earth observation, hydrology, meteorology, earthquake monitoring, and geography to health and economy. Nevertheless, the need for shared standards to process and analyse different representations of data modelling remains a priority to support comprehensive risk management and flows of information about financing. Linking the indicators of the New Ten Essentials – Disaster Resilience Scorecard of the United Nations Office for Disaster Risk Reduction (UNDRR) at the local level and the Sendai Framework for the Disaster Risk Reduction (SFDRR) National Monitoring System using an open data platform can support local governments in integrating resilience action plans into national disaster monitoring systems by including the component for monitoring displacement caused by disaster. This allows for developing comparable and coherent data sets that can be "sufficiently consistent and comparable to allow meaningful measurement of progress and impact" (UNDRR, 2017).

Considering the limitation of access for the most vulnerable population (IDPs and refugees) because of political and security restrictions beyond local governments in the fragile context of MENA, this can be followed with a cross-sectional comparison of the results assessing the resilience of the different groups of participants in public consultations to develop a comprehensive action plan for city resilience. This step can then be followed with a validation process of the proposed action plans for resilience with all groups to ensure accommodating the voices of all DRR stakeholders equally, "leaving no one behind" (Menash *et al.*, 2020).

The need for shared standards to process and analyse different representations of data modelling remains a priority to support Open Disaster Displacement Data management and information flows for building resilience in fragile and conflictual settings.

It is important to emphasise the limitations of using open data as a mechanism for sharing knowledge and raising awareness of risk because, in grassroots-level communities, social networks might work as a more efficient means for knowledge-sharing, especially when the level of literacy, access to technology, and useability of web-based mobile devices is limited. This claim is supported by the experience shared by a city authority representative from MENA:

> The use of social networks can enable the passing of information and intervention to people in a way that is acceptable to them. We realised we use a lot of modern conventional techniques working through workshops, seminars and exchange visits (usual jargon) and working through the chief administration government structures, other times we spend a lot of money on this, but we are back to square one. One of the studies we are applying now is the use of social networks to influence the behaviour of change and adaptation mechanisms. These relationships are very important because they make things work. It is a combination of techniques, weigh them out and see what works best. What works in one context may not necessarily be suitable in other.

We have no problem with data collection, we have a problem in data updates, data integration (connectivity between different agencies ex: Ministry of Infrastructure and MET Office) data ownership and sharing (Ministry of Infrastructure, 2018).

In this chapter, it is concluded that international frameworks and tools for assessing resilience overlook the sub-regional challenges that are experienced historically in the MENA Region and similar fragile settings globally. A bottom-up, custom-based adaptation to climate change and innovative mechanisms to enhance the commitments of national government to the 2015–2030 Global Agendas and to shift policy legislation and implementation for displacement caused by disasters and conflict from "soft-laws" to "hard-laws is required in these settings. Similarly, there is a need to overhaul current tools for assessing resilience and reporting mechanisms for global frameworks within the local context of fragility in MENA cities, to improve the accuracy, transparency and validity of the data required to structure effective and efficient action plans, and monitoring for urban resilience to disaster, "leaving no one behind" (Mensah, *et al.*, 2022). This will help decision-makers to enhance societal resilience, complementing the matrix of data about loss caused by disaster with human security and displacement caused by conflict in the MENA Region, to guide the approach of future research that can be adapted to similar contexts world-wide.

Reference list

Brooks, N. 2003. Vulnerability, risk and adaptation: A conceptual framework. *Working Paper 38*, pp. 1–16. Tyndall Centre for Climate Change Research.

Burger, M., & Wentz, J. A. 2015. *Climate Change and Human Rights*. United Nations Environmental Programme.

Caretta, A., Mukherji, M. A., Betts, R. A., Arfanuzzaman, M. S., Morgan, M. R., & Kumar, M. 2022. Water. In: *Climate Change 2022: Impacts, Adaptation, and Vulnerability. Contribution of Working Group II to the Sixth Assessment Report of the Intergovernmental Panel on Climate Change*. Available online: https://www.ipcc.ch/report/ar6/wg2/

Cheema, S. 2022. *UNDP and the Democratic Governance Agenda. Handbook on Governance and Development*, p. 340. Available Online: https://doi.org/10.4337/9781789908756

Crichton, D., 1999. The risk triangle. *Natural disaster management*, 102(3): 102–103.

Cutter, S. L., & Gall, M. 2015. Sendai targets at risk. *Nature Climate Change*, 5(8): 707–709.

Feldmeyer, D., Wilden, D., Kind, C., Kaiser, T., Goldschmidt, R., Diller, C., & Birkmann, J. 2019. Indicators for monitoring urban climate change resilience and adaptation. *Sustainability*, 11(10): 2931.

Florin, M., & Bürkler, M. T. 2017. *Introduction to the IRGC Risk Governance Framework. No. REP_WORK*. EPFL.

Gall, M. 2013. *From Social Vulnerability to Resilience: Measuring Progress Toward Disaster Risk Reduction*. Unu-Ehs.

Mensah, H., & Ahadzie, D. K. 2020. Causes, impacts and coping strategies of floods in Ghana: a systematic review. *SN Applied Sciences*, 2: 1–13.

Mensah, J., Mensah, A., & Mensah, A. N. 2022. Understanding and promoting the 'leaving no one behind' ambition regarding the sustainable development agenda: A review. *Visegrad Journal on Bioeconomy and Sustainable Development*, 11(1): 6–15.

Pörtner, H. O., Roberts, D. C., Adams, H., Adler, C., Aldunce, P., Ali, E., Begum, R. A., Betts, R., Kerr, R. B., Biesbroek, R., & Birkmann, J. 2022. *Climate change 2022: impacts, adaptation and vulnerability*. IPCC.

Schreyer, P. 2012. Output, outcome, and quality adjustment in measuring health and education services. *Review of Income and Wealth*, 58(2): 257–278.

Shaffer, G. C., & Pollack, M. A. 2009. Hard vs soft law: Alternatives, complements, and antagonists in international governance. *Minnesota Law Review*, 94: 706.

Stein, S., & Walch, C. 2017. The Sendai framework for disaster risk reduction as a tool for conflict prevention. In: *Conflict Prevention and Peace Forum*. Social Science Research Council.

United Nations Environment Programme (UNEP). 2015. *Early Warning as a Human Right: Building Resilience to Climate-Related Hazards*. Available at: https://wedocs.unep.org/handle/20.500.11822/7429

United Nations Office for Disaster Risk Reduction (UNDRR). 2017. *Sendai Framework data readiness review – Global summary report*. Available from: https://www.unisdr.org/we/inform/publications/53080

United Nations Office for Disaster Risk Reduction (UNDRR). 2022. *Comprehensive Disaster and Climate Risk Management*. Available at: www.undrr.org/comprehensive-disaster-and-climate-risk-management-crm

United Nations Office for Disaster Risk Reduction (UNDRR). 2023a. *Mid-term Review of the Sendai Framework*. Available at: https://sendaiframework-mtr.undrr.org/about-sendai-framework-midterm-review

United Nations Office for Disaster Risk Reduction (UNDRR). 2023b. Sendai framework for disaster risk reduction goal, targets and metrics. *Sixth Workshop under the Glasgow – Sharm el-Sheikh Programme on the Global Goal on Adaptation: Target-Setting, Metrics, Methodologies and Indicators*. Available at: https://unfccc.int/sites/default/files/resource/UNDRR%20Presentation_Sendai%20Framework-GGA6%20%281%29.pdf

United Nations Office for Disaster Risk Reduction and World Meteorological Organisation (UNDRR and WMO). 2022. Global status of multi-hazard early warning systems. *Target G. Sendai Framework for Disaster Risk Reduction*. Available Online: http://www.undrr.org/publication/global-status-multi-hazard-early-warning-systems-target-g

Watanabe, K. 2021. *Practical methods for DRR investment acceleration through organizing DRR strategy in the local*. Available Online: https://www.undrr.org/media/73551/download?startDownload=true

World Meteorological Organisation. 2021. *Weather-Related Disasters Increase Over Past 50 Years*. Available at: https://public.wmo.int/en/media/press-release/weather-related-disasters-increase-over-past-50-years-causing-more-damage-fewer

World Meteorological Organisation. 2022. *Early Warnings for All*. WMO. Available at: https://public.wmo.int/en/earlywarningsforall

6   Adaptation to Climate Change – Financing Urban Resilience

The Gap in Financing for Urban Resilience and Climate Change Adaptation (CCA)

The economic case for investing in adaptation to climate change is powerful, with cost-benefit ratios ranging from 1:2 to 1:10 (GCA, 2023). In the 2020 United Nations Environment Programme (UNEP) Adaptation Gap Report, it is estimated that finance for adaptation represented only 14% of total public finance, and rates of issuance of green bonds globally for adaptation purposes were low (UNEP, 2021). An early response to adaptation leads to a "triple dividend" of avoided losses, economic benefits, and benefits to society and the environment (Bharadwaj and Shakya, 2021). However, in a fragile setting, since violent and armed conflict can lead to the destruction and sometimes wilful loss of productive assets for Climate Change Adaptation (CCA) and increase the risk of loss of investments, this requires the development of innovative mechanisms for mainstreaming climate-related security risks into finance for CCA to help to leverage the co-benefits of climate-related action for peace and security (UNDP, 2022).

In the recent 2017 UNDRR guide on DRR terminologies, "risk" is defined in the context of financial losses as

> risk transfer in the process of formally or informally shifting the financial consequences of particular risks from one party to another, whereby a household, community, enterprise or state authority will obtain resources from the other party after a disaster occurs, in exchange for ongoing or compensatory social or financial benefits provided to that other party.
>
> (UNDRR, 2017a)

This definition is useful for quantifying post-disaster losses but ignores the principles of risk mitigation and prevention, especially in the context of fragile cities where the risk of violent conflict can be forecasted in parallel to early warning systems for risk of disasters and to help the best management of institutional capacities and investments for emergency response and long-term recovery.

DOI: 10.4324/9781003363224-6

Top-down and high-level investments in urban resilience have been widely divergent among governments, international organisations, and prominent institutions. Owing to the uncertainty associated with climate change scenarios, the lack of understanding of adaptation to climate change, local engineering, operational, and cost realities, and the generalising nature of top-down modelling itself, such estimates are likely to remain inconclusive for years to come (OECD, 2019). According to the top-down models, it is agreed that infrastructure and urban areas will bear the majority share of future adaptation costs, despite their variances in cost estimates. Accordingly, "resilience investment opportunities" (ADB, 2013) represent a large proportion of finance requirements for urban adaptation. It is an opportunity to increase the resilience of urban assets, areas, and systems through profitable, market-based investments that reduce risks.

In the situation of disaster, armed conflict, internal displacement, and displacement across borders, the absence of rules and principles for humanitarian assistance can lead to prioritising investments in emergency response, ignoring the long-term demand for disaster risk reduction. Evidence from previous literature indicates that a sizeable gap exists between investments in disaster resilience and spending for conventional response to crises. According to some estimates, for every $100 spent in development aid, only 40 cents has been invested in reducing the impact of disasters. At the same time, losses caused by disaster in developing nations amount to $862 billion (a considerable under-estimate) – equivalent in value to one-third of all international development aid (UNDRR, 2017).

Taking into account the differences in fiscal resources and borrowing capacities that exist throughout MENA countries, assessing the adequacy of different financial and budgetary tools for urban resilience will be useful to accommodate climate change emergencies of varying frequency and severity. Integrating assessments of fiscal risk that take climate into account with concepts assessing urban resilience, introduced in Chapter 4, will help to "provide a sound basis for targeting investments in risk reduction and adaptation and developing a strategy to ensure adequate funding for recovery and building back better" (OECD, 2022).

Assessments of Climate Risks and Fiscal Risk Management

When addressing the gaps in investments in adaptation and resilience to climate change, it is important to understand how finance is mobilised for adaptation and investigate the role of data in assessing climate risks to create an enabling environment for financing. Firstly, it is essential to acknowledge and comprehend the components of climate hazards, exposure, vulnerability, and sources of physical and transition risks associated with climate change. "This includes both data about past risks, disclosures about current risks and projections about future ones" (OECD, 2022). Noting the impact of residual physical risks that can translate into fiscal and financial risks once absorbed by the economy, providing standardised and verified data about climate assessments can help to guide incentives for risk reduction across all society sectors (individuals, corporations, the financial industry, and local governments) to be incorporated into frameworks for urban resilience and public financial management for CCA.

The need for the architecture of data to support mobilising finance is critical to help to fill the gap in current data that lack consistency and transparency, noting that "monitoring the performance and scalability of adaptation finance remains difficult due to challenges associated with context dependency, confidentiality restrictions, uncertain causality, and a lack of unified and agreed-upon impact metrics" (TCFD, 2022). In the context of the MENA Region, having a regional, open baseline dataset for climate can help to internalise the externalities, build investment markets, and develop new types of financial products and KPIs for sustainability-linked bonds. Investments in data about climate risk will help to create new opportunities towards a common data architecture and open risk metrics that will help to align finance with climate-resilient development goals.

Another important aspect to be considered is the regular reporting of climate-related fiscal risks. This can help to increase public awareness and capabilities of government institutions in developing resilience-related financial products and providing guidance on the role of private investment in financing resilience to climate. Promoting transparency in public financial management, reporting of climate-related financial information, and financial disclosure regarding climate can all help to guide strategic planning for budget allocation and prioritisation of investment in humanitarian aid, while evaluating risks and exposures over the short, medium, and long terms, and "strengthen confidence from financial markets that countries can manage the impact of chronic and acute climate change (e.g., extreme weather events)" (OECD, 2022).

Considering the challenges of risk governance and municipal budgeting in the fragile settings of the MENA Region, financial disclosure regarding climate should be integrated into the strategy for fiscal risk management. In the recent United Nations Development Programme Report, several actions were outlined that can help to leverage the co-benefits of climate action for peace and security and how the implementation of finance to address climate can interact with drivers of fragility and insecurity (UNDP, 2021b).

Examining the amount of $14 billion to address climate implemented in 146 countries, including 46 fragile contexts, over the period 2014–2021, it was indicated in the report that allocating finance to address climate cannot ignore conflict and fragility. "Access to climate finance means, ensuring climate finance reaches the last mile to support the most vulnerable contexts", otherwise finance will increase vulnerabilities and widen the gap, exacerbating climate-related security risks (UNDP, 2021).

In fragile contexts where non-state armed groups have stronger influence, security risks associated with safe access to operational services can be higher than investment costs and, in some cases, aggravate tensions or conflict dynamics. Thus, regular situational assessments to decide when to invest and geopolitical analysis to decide where to invest are critical in fragile settings to reduce risks and increase the profitability of private and public financing. This can vary between climate adaptation and mitigation (as shown in Figure 6.1.)

According to the OECD, of total public climate finance, 70 percent targeted mitigation and just 21 percent, adaptation, with the remainder, cross-cutting. As for private sector finance mobilised by developed countries, 93 percent focused on mitigation, primarily the energy sector (60 percent) and in middle-income countries.

(OECD, 2020)

Scenario analysis can be utilised in disclosure of climate-related risks and opportunities as a mechanism to help to develop hypothetical constructs and predictions of situational sensitivity and uncertainty in fragile settings and to explore alternatives that might significantly alter the basis for "business-as-usual" assumptions (TCFD, 2017). Considering the intersections between disasters caused by climate change and risks of damage, destruction, and displacement associated with conflict, multiple scenarios should be used in the context of the MENA Region to explore how different permutations and/or temporal developments of the same key factors can yield very different outcomes (Verner, 2012).

To ensure a project does not entrench existing grievances or widen social differentiation, adaptation planners and implementers should be encouraged to ask questions such as: how would the planned intervention affect the livelihoods of different groups, or their access to resources? Could it exacerbate the exclusion of specific communities? Could its outcomes be perceived as unfair, benefiting one group more than another?

(Remling, 2022)

In this instance, capturing experiences of key role-players on the ground is critical, using both quantitative and qualitative methods to illustrate potential pathways and outcomes and to ensure that inclusive reflection of the needs and priorities of all beneficiaries is considered in the decision-making process. Further investigation of the identity of local community and governmental key stakeholders, donors, and development partners who are operating in the field of Climate Change Adaptation (CCA) and Disaster Risk Reduction (DRR) and how best to engage them in Urban Resilience Governance to enhance financing to address climate is explored in Chapter 7.

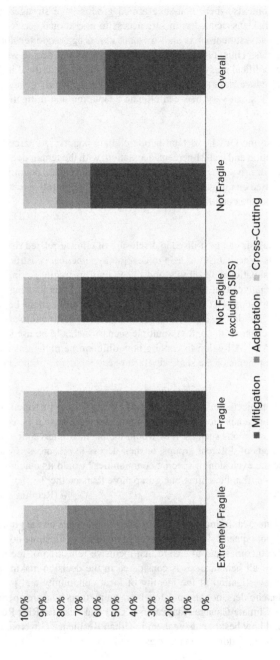

Figure 6.1 Total vertical fund financing for mitigation, adaptation, and cross-cutting priorities by country fragility classification 2014–May 2021 (not including co-financing) (UNDP, 2021)

There is limited evidence to date that programmes for CCA are being implemented in a conflict-sensitive manner (Peters *et al.*, 2020). Exploring the disregard of conflict in allocating finance to address adaptation to climate change is another aspect that is critical to consider in the fragile and conflictual settings in MENA, where maladaptation can have a negative impact on conflict resolution and peacebuilding efforts. Ill-designed adaptation programmes can precipitate grievances and conflict, causing unintended harm (Alcayna, 2020), while increasing environmental degradation, loss of bio-diversity, increasing insecurity of land tenure, and the marginalisation of minority groups.

The "holistic understanding of the two-way interaction between interventions and the conflict context (i.e., the intervention's influence on conflict, and how the conflict contexts affect the intervention)" is one of the key principles investigated in the Supporting Pastoralism and Agriculture in Recurrent and Protracted Crises SPARC (2021) Synthesis Report, in which it is indicated that "peaceful aims do not always guarantee peaceful outcomes and some well-meaning development programmes actually end up fuelling conflict" (Cited in SaferWorld, 2008). Considered in the context of MENA, this can be associated strongly with DRR governance structures and the operational mechanisms of directing and controlling funding for CCA where national policies and institutional powers can be exploited by the elite groups and politically dominant members of society to derive personal gains from public profits and reduce productivity of investments to address climate which, in turn, lowers economic growth rates (IMF, 1998), distorts markets, undermines competition, and leads to missed business opportunities and private sector investments.

Adaptive, Absorptive, and Transformative Capacity Building

Vulnerability varies across time and space with the variations in the vulnerabilities of displaced people and the adaptive capacity of local governments (hosting communities). Considering the view of Harris *et al.* (2013) that "vulnerability is dynamic and shaped by interconnected shocks and stresses, and how it must be addressed as such", accessibility to humanitarian aid and the institutional capacity of local government to pursue progress into early recovery and long-term development can also be restrained by the state of political and economic complexities and fragile conflict settings.

The adaptive, absorptive, and transformative capacities of local governments in managing financing for urban resilience and adaptation to climate change also plays a critical role in directing investments towards the necessary channels while maintaining accountability for policy formulation and the performance of solutions implemented. Considering the fragile context of the MENA Region, the fragmentation of local government administration, misalignment in fiscal budgeting between different levels of government (local, regional, national, and transnational), institutional political inertia, and weak administrative practices all result in widening the gap of the technical capacities of local governments in "analysing, explaining and communicating climate change risks, costs and benefits of adaptation amongst investors and public authorities" (TCFD, 2022).

In terms of technical capacity, first defined by Béné *et al.* (2012) as "the ability of a system to adjust, modify or change its characteristics and actions to moderate potential future damage and to take advantage of opportunities", local communities and city authorities in the MENA Region cannot continue to function without major changes in functional and structural identity to enhance its adaptive capacities and develop the expertise of action planning for resilience to develop robust business models, investment opportunities suited to external financiers, and match adaptation needs. Thus, financial investments in the diversification of urban livelihoods through public-private partnerships can play a vital role in delivering basic services and introducing early response and long-term resilience to measures for CCA.

Evidence from the recent IPCC report indicates that, despite the increase in flows of finance to address climate, disbursements and investments are lagging behind in comparison with needs. "Sustainable finance is a crucial tool to support economic growth while reducing environmental pressures and considering social and governance aspects" (EC, 2022). Thus, sustainable financial investments and developing coping mechanisms to be introduced during periods of shock are key to enhancing absorptive capacities. Defined as "the ability of a system to prepare for, mitigate or prevent negative impacts, using predetermined coping responses in order to preserve and restore essential basic structures and functions", absorptive capacity could help to deliver key outcomes such as adjustments in the cost of capital, adjustment in liquidity, de-risking, larger capital flows for CCA and investments in urban resilience (Béné *et al.*, 2012).

In all national climate policies (NAPA, National Biodiversity Action Plan, NDC, NAP), the strong link between conflict resolution and climate adaptation is recognised (Cao *et al.*, 2021). With the focus on rural settings, agricultural, and pastoral communities, it is important to address the impact of climate policies in urban settings and how to advance climate investments in cities to build resilience for the most vulnerable communities of IDPs and refugees. Some nations in MENA have started the National Adaptation Plan (NAP) and procedures for sectoral climate change strategy, and they are expanding on those current plans to advance the NAP process. Nevertheless, coordination between horizontal and vertical levels and the implementation of long-term action plans for adaptation, including the mobilisation of financial resources, budgeting process integration, and enhancement of technical capacities, all remain challenges to be addressed through building institutional capacities to mainstream adaptation to climate (UNDP, 2019).

The status above requires moving from enhancing only adaptive capacities to climate change to understanding the need for building absorptive and transformative capacities within institutional governance systems in fragile settings as a key to help to sustain climate investment and direct its co-benefits into local communities and economic growth (International Monetary Fund, 1998). Absorptive capacity is defined as "the ability of a system to prepare for, mitigate or prevent negative impacts, using predetermined coping responses in order to preserve and restore essential basic structures and functions" (Béné *et al.*, 2012). This includes coping mechanisms used during periods of shock, which can help to delay debt repayments and reduce insurance premiums. On the other hand, building transformative capacities to enhance climate investments becomes a necessity in the MENA context with "the ability to create a fundamentally new system so that the shock will no longer have any impact. This includes the introduction of conflict resolution mechanisms, urban planning measures, and actions to stamp out corruption" (Béné *et al.*, 2012). An understanding of the needs for building transformative capacities to address climate change in fragile settings, such as "the introduction of conflict resolution mechanisms, urban planning measures, and actions to stamp out corruption" (Béné *et al.*, 2012), can enhance economic competitiveness, financial stability, social cohesion, and the co-benefits of climate policies.

Based on this chapter, it is concluded that to enhance financing to address climate in fragile settings, greater conflict sensitivity is key to avoid maladaptation and increase co-benefits.

> Exercising greater conflict sensitivity, including a broader understanding of the impacts of climate and non-climate induced conflict and security risks on climate action, can improve risk management. Equally, qualification of co-benefits or peace dividends can incentivise much-needed investments in fragile and conflict-affected contexts, the most severely affected of which are among those that have the least access to climate finance.
>
> (UNDP, 2022)

Reference list

Alcayna, T. 2020. *At What Cost: How Chronic Gaps in Adaptation Finance Expose the World's Poorest People to Climate Chaos.* Available Online: https://reliefweb. int/report/world/what-cost-how-chronic-gaps-adaptation-finance-expose-worlds-poorest-people-climate

Asian Development Bank (ADB). 2013. *Investing in Resilience: Ensuring a Disaster-Resistant Future.* Asian Development Bank. Available Online: https://www.adb.org/ publications/investing-resilience-ensuring-disaster-resistant-future

Béné, C., Wood, R. G., Newsham, A., & Davies, M. 2012. Resilience: New utopia or new tyranny? Reflection about the potentials and limits of the concept of resilience in relation to vulnerability reduction programmes. *IDS Working Papers 2012, 405,* pp. 1–61. Institute of Development Studies.

Bharadwaj, R., & Shakya, C. (Eds.). 2021. *Loss and Damage Case Studies from the Frontline: A Resource to Support Practice and Policy*. International Institute for Environment and Development (IIED).

Cao, Y., Alcayna, T., Quevedo, A., & Jarvie, J. 2021. *Exploring the Conflict Blind Spots in Climate Adaptation Finance. Synthesis Report*. Overseas Development Institute.

European Commission (EC). 2022. *Overview of Sustainable Finance*. Available at: https://finance.ec.europa.eu/sustainable-finance/overview-sustainable-finance_en

Global Centre on Adaptation (GCA). 2023. *Adaptation Finance*. Available at: https://gca.org/programs/climate-finance/

Harris, K., Keen, D., & Mitchell, T. 2013. *When Disasters and Conflicts Collide: Improving Links Between Disaster Resilience and Conflict Prevention*. ODI.

International Monetary Fund (IMF). 1998. *Roads to Nowhere: How Corruption in Public Investment Hurts Growth*.

Organisation for Economic Co-operation and Development (OECD). 2019. Implementing adaptation policies: towards sustainable development. Issue Brief. Available Online: https://www.oecd.org/g20/summits/osaka/OECD-G20%20Paper-Adaptation-and-resilient-infrastructure.pdf

Organisation for Economic Co-operation and Development (OECD). 2020. *Climate Finance Provided and Mobilised by Developed Countries in 2013–18 Key Highlights*. Available at: www.oecd.org/environment/cc/Key-Highlights-Climate-Finance-Provided-and-Mobilised-by-Developed-Countries-in-2013-18.pdf

Organisation for Economic Co-operation and Development. 2022. *Building Financial Resilience to Climate Impacts: A Framework for Governments to Manage the Risks of Losses and Damages*. OECD Publishing. https://doi.org/10.1787/9e2e1412-en.

Peters, K., Dupar, M., Opitz-Stapleton, S., Lovell, E., Budimir, M., Brown, S., & Cao, Y. 2020. *Climate Change, Conflict and Fragility: Information and Analysis to Support Programme Design Scoping for the Climate and Resilience Framework Programme (CLARE)*. Available Online: https://odi.org/en/publications/climate-change-conflict-and-fragility-an-evidence-review-and-recommendations-for-research-and-action/

Remling, E. 2022. *Five Rules for Climate Adaptation in Fragile and Conflict-Affected Situations*. Avilable Online: https://www.developmentresearch.eu/?p=1194

SaferWorld. 2008. *Conflict-Sensitive Development*. Available at: www.saferworld.org.uk/resources/publications/313-conflict-sensitive-development

Task Force on Climate-related Financial Disclosures (TCFD). 2017. *The Use of Scenario Analysis in Disclosure of Climate-related Risks and Opportunities*. Available at: www.tcfdhub.org/scenario-analysis/

Task Force on Climate-related Financial Disclosures (TCFD). 2022. *Climate Change Presents Financial Risk to the Global Economy*. Available at: www.fsb-tcfd.org/

United Nations Development Programme (UNDP). 2017. *Regional Briefing on National Adaptation Plans. Middle East and North Africa in Focus*. Available at: www.adaptation-undp.org/sites/default/files/resources/regional_briefing_on_naps_mena_in_focus.pdf

United Nations Development Programme (UNDP). 2019. United Nations Development Programme Background Guide. National Model United Nations. Available Online: https://www.nmun.org/assets/documents/conference-archives/new-york/2019/ny19-bgg-undp.pdf

United Nations Development Programme (UNDP). 2021a. *Making Climate Finance Work for Conflict-Affected and Fragile Contexts*. Available at: www.undp.org/

publications/climate-finance-sustaining-peace-making-climate-finance-work-con-flict-affected-and-fragile-contexts

United Nations Development Programme (UNDP). 2021b. *Nationally Determined Contributions (NDC). Global Outlook Report 2021. The State of Climate Ambition.* Available at: www.undp.org/sites/g/files/zskgke326/files/2021-11/UNDP-NDC-Global-Outlook-Report-2021-The-State-of-Climate-Ambition.pdf

United Nations Development Programme (UNDP). 2022. *How Can Climate Finance Work Better for Fragile and Conflict-Affected Regions?* Available at: www.undp.org/blog/how-can-climate-finance-work-better-fragile-and-conflict-affected-regions

United Nations Environment Programme. 2021. *Adaptation Gap Report 2020.* Nairobi.

United Nations Office for Disaster Risk Reduction (UNDRR). 2017a. *Financing Prevention.* Geneva. Available at: www.undrr.org/financing-prevention

United Nations Office for Disaster Risk Reduction (UNDRR). 2017b. *Sendai Framework Terminology on Disaster Risk Reduction.* Geneva. Available at: www.undrr.org/terminology

Verner, D. (Ed.). 2012. *Adaptation to a Changing Climate in the Arab Countries: A Case for Adaptation Governance and Leadership in Building Climate Resilience. No. 79.* World Bank Publications.

7 Urban Resilience Governance – Key Stakeholders in Climate Change Adaptation (CCA) and Disaster Risk Reduction (DRR)

Definition of Key Stakeholders – Characteristics, Roles, and Responsibilities

Since action plans for urban resilience are increasingly reliant on the voluntary effort and ownership of DRR stakeholder bodies and are affected by their organisational powers, it is important to identify the parameters for understanding risk and assessing resilience from the perspective of DRR key stakeholders, develop effective and inclusive operational programs for their engagement in the decision-making process, and form legislative policy guidelines beyond the theoretically-bound indicators of resilience to disaster and numerically generated indices.

In Stakeholders Theory, which dates back to 1984 (Freeman, 1984), stakeholders are defined as "any group or individual who can affect or is affected by the achievement of the organisation's objectives". The extent to which the stakeholders can affect or are affected by the process of assessing resilience to disaster and the decision-making process for action plans cannot be determined without the "identification, classification, analysis, and management" of the strategy to engage stakeholders (Cleland, 1986, Littau et al., 2010). From a theoretical perspective of project management, the development of a disaster resilience action-plan is based on two main project management processes: project evaluation (assessing the existing gaps in mechanisms for understanding and implementing resilience) and project strategy (developing policies to identify priorities for investment and action) (Freeman et al., 2010). Project management theories were investigated in this study to identify the qualitative variables of DRR key stakeholders and quantitative attributes required to measure their level of involvement in the process of assessing resilience to disaster, taking into account the effectiveness of their engagement in the development of action plans for city resilience (Littau et al., 2010).

In the literature review "25 Years of Stakeholder Theory in Project Management Literature", Littau et al. (2009) classified 116 articles using the term "stakeholder". The classification process was based on the percentage of articles about stakeholders in published journals and sources

DOI: 10.4324/9781003363224-7

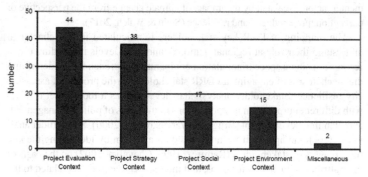

Figure 7.1 Allocation of articles by context. 25 Years of Stakeholder Theory in Project Management Literature (1984–2009) (Littau *et al.*, 2009)

of information (reviews case studies empirical data), and the origin of the articles by country industry sector such as general, construction, process industries, manufacturing, information and services, agriculture/development, facilities, and utilities.

Focusing on the scope of this study, theory about the context of stakeholder engagement highlighted the context of project evaluation, combining the categories of project risk and project performance. Based on the study by Littau *et al.* (2009), the highest percentage of 44% for project evaluation was scored for articles allocated by context, compared with social and environmental contexts of project strategy, as shown in Figure 7.1. This showed the importance of the role of stakeholders in project evaluation and the need to define well their level of engagement and responsibilities in evaluating progress on building resilience for disaster risk reduction (Littau *et al.*, 2009).

Mapping DRR and CCA Key Stakeholders

In paragraph 48 (c) of the Sendai Framework, the United Nations Office for Disaster Risk Reduction (UNDRR) is called upon "in particular, to support the implementation, follow-up and review of this framework through generating evidence-based and practical guidance for implementation in close collaboration with States, and through mobilisation of experts; reinforcing a culture of prevention in relevant stakeholders" (UNDRR, 2015). However, without taking into account the underlying socio-economic drivers of disaster and conflict risk evident in the MENA Region, developing a number of Sendai Framework implementation guides, targeting the region's context of poverty, inequality, and marginalisation associated with socio-political and institutional factors of conflict and disaster to strengthen disaster risk governance at the local, national, regional, and global level will play a strong role in peacebuilding, while mitigating the environmental factors escalated by the impact of climate change from the perspective of prevention, preparedness, and resilience (Stein & Walch, 2017).

The mapping of DRR key stakeholders is articulated in this chapter to understand their role at regional, national, and local levels in building urban resilience using a space triangulation comparative approach while identifying the mechanisms of engaging all DRR stakeholders in the process of assessing city resilience. Stakeholder analysis has developed as a tool, or set of tools with different purposes, in its applications in the fields of policy, management, and planning development (Brugha & Varvasovszky, 2000). Stakeholder analysis refers to an approach for understanding a system by identifying the key role=players or stakeholders (Ramirez, 1999; Chevalier, 2002) on the basis of their attributes, inter-relationships, and their respective interests related to the system. Stakeholder analysis is a central theme in conflict management and dispute resolution (Swiderska *et al.*, 2002), which is important to explore in the region's fragile and conflictual settings. Considering that migrants were only noted in the 2012 Rio+20 Outcome Document together with local communities, these were not included in the nine categories of DRR key stakeholders (theoretical population) officially listed by the UNDRR in 2018 as part of the Strategic Approach to Capacity Development for the implementation of the Sendai Framework for Disaster Risk Reduction as follows:

1 National Government (including elected leaders, parliamentarians, and line ministries)
2 Local and Sub-National Government
3 Private Sector and Professional Organisations
4 Non-Governmental and Civil-Society Organisations (NGOs and CSOs)
5 Education and Research Institutions
6 Individuals and Households
7 Media
8 Regional Organisations including IGOs
9 The UN, International Organizations (IGOs), and International Financial Institutions (IFIs)

In February 2018, this categorisation was expanded by the UNDRR Stakeholder Advisory Group and the establishment of whole-of-society engagement strategy, as shown in Figure 7.2. Following two days of discussions, a new list was approved including 50 representatives from the private sector, trade unions, farmers, education, community-based groups, NGOs, indigenous people groups, and the media (UNDRR, 2018). This list was formed to support the implementation process of the SFDRR global priorities (ADRC, 2015), reduce disaster losses, and build coherence with the 2015–2030 Global Agendas for the Paris Agreement on Climate Change, the Sustainable Development Goals, and Habitat III New Urban Agenda (Murray *et al.*, 2017).

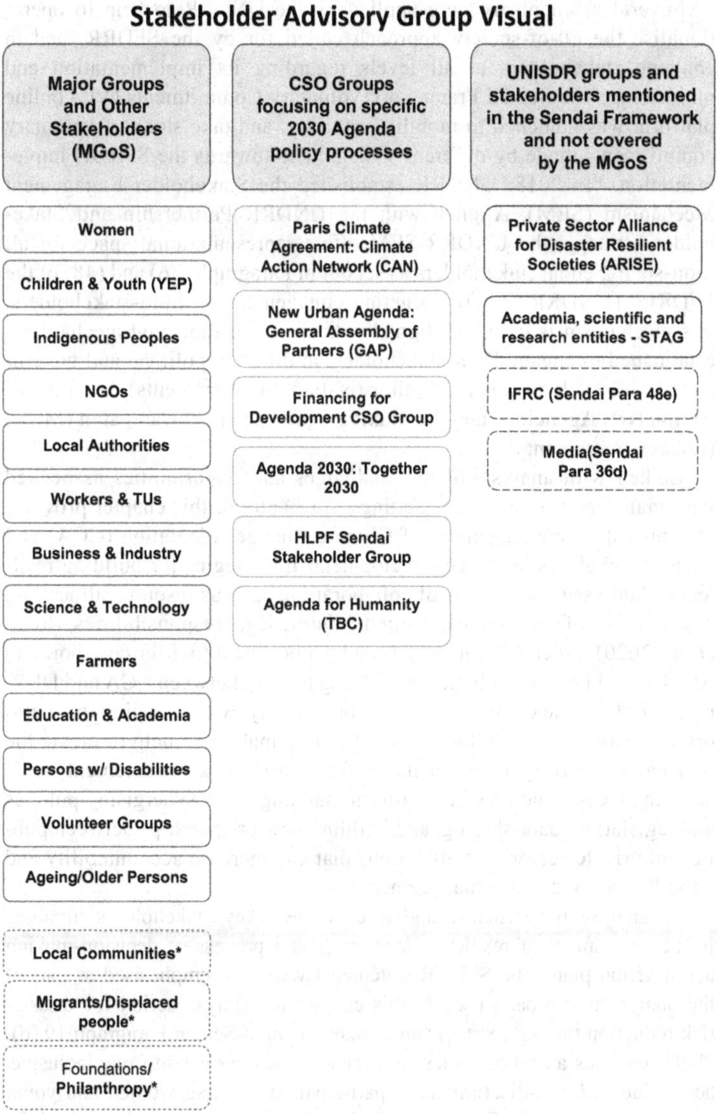

Stakeholder Advisory Group Visual

Major Groups and Other Stakeholders (MGoS)	CSO Groups focusing on specific 2030 Agenda policy processes	UNISDR groups and stakeholders mentioned in the Sendai Framework and not covered by the MGoS
Women	Paris Climate Agreement: Climate Action Network (CAN)	Private Sector Alliance for Disaster Resilient Societies (ARISE)
Children & Youth (YEP)	New Urban Agenda: General Assembly of Partners (GAP)	Academia, scientific and research entities - STAG
Indigenous Peoples	Financing for Development CSO Group	IFRC (Sendai Para 48e)
NGOs	Agenda 2030: Together 2030	Media(Sendai Para 36d)
Local Authorities	HLPF Sendai Stakeholder Group	
Workers & TUs	Agenda for Humanity (TBC)	
Business & Industry		
Science & Technology		
Farmers		
Education & Academia		
Persons w/ Disabilities		
Volunteer Groups		
Ageing/Older Persons		
Local Communities*		
Migrants/Displaced people*		
Foundations/ Philanthropy*		

Figure 7.2 UNDRR Stakeholder Advisory Group (UNDRR, 2018)

Several mechanisms were applied by the UNDRR to help to operationalise the all-of-society approach called for by the SFDRR, and to convene stakeholders at all levels regarding its implementation and monitoring. The Sendai Framework Voluntary Commitments (VC) online platform was launched to mobilise, monitor, and take stock of voluntary commitments made by different stakeholders towards the SFRDD implementation. "In 2018, UNDRR established the Stakeholder Engagement Mechanism (SEM). Aligned with the UNDRR Partnership and Stakeholder Strategy, the UNDRR-SEM offers representational space for all 'non-state' Sendai stakeholders as set out in Paragraph (36) and (48) of the SFDRR" (UNDRR, 2019). Targeting engagement of multi-stakeholders, this mechanism is strongly aligned with the other tools and mechanisms established to engage DRR stakeholders at different political and governance levels such as Sendai Monitor (National Governments), UN Plan of Action (UN Agencies), and the Making Cities Resilient Campaign (MCR) (Local Governments).

Linked with analysis of the challenges and opportunities associated with maladaptation in fragile settings in Chapter 6, this chapter provides insights into the engagement of Climate Change Adaptation (CCA) and DRR stakeholders in the co-development of strategies for building resilience. Addressing key areas of collaboration, such as resource allocation, effectiveness of communication, and defining legal responsibilities, Booth *et al.* (2020) indicated that media can be a powerful tool for collaborators which could be utilised better in bridging the gap between CCA and DRR, particularly in the communication of uncertainty, while keeping the memory of disasters alive in the minds of policy-makers to help to create the transparent delivery of mechanisms for disaster risk management while adopting CCA principles in long-term planning. New synergising policies and legislation, data-sharing, and building new partnerships between public and private sectors are all actions that can increase accountability and trans-boundary disaster management.

To analyse the structure and role of DRR key stakeholders involved in the assessment of resilience to disaster and process of decision-making about action plans, the SFDRR statements were first emphasised as part of the analytical approach used in this chapter to catalyse action for disaster risk reduction through partners and stakeholders in Sendai Paragraph 19 (d): "DRR requires an all-of-society engagement and partnership . . . inclusive, accessible and non-discriminatory participation . . . and women and youth leadership". In Sendai Paragraph 35, the role of informal actors beyond the institutionally defined DRR stakeholders is highlighted, "while states have the overall responsibility for reducing disaster risk, it is a shared responsibility between governments and relevant stakeholders. Non-state stakeholders play an important role as enablers in providing support to states". Further guidance was captured from the opinion of ODI expert and co-author of *Supporting governance for climate resilience: Working with political institutions* (Fraser & Kirbyshire, 2017), as stated:

> Stakeholders mapping, and engagement processes are introduced, who they are, how they engage, and looking into international donors' short-term technical input, the accountability of governments and donors, the distribution of knowledge, resources, the internal politics and institutional mechanisms that shall undermine resilience. The role and influence of different political parties and relations of national and local government stakeholders are all issues we do not know a lot about. How the discussions on resilience unfold are only available as theoretical perspectives, and a more adaptive learning model is required.
>
> (ODI, 2017)

Dynamics of Key Stakeholder Engagement

In addressing the dynamics of key stakeholder engagement, it is recalled in this chapter that one of the main limitations assessing resilience is gaining access to civil society organisations representing the IDPs and refugees in the fragile context of conflict and displacement in the MENA Region. This is maintained to highlight the impact that civil society organisations can have in mitigating disputes and resolving conflicts. Assal (2016) emphasises that "the role of civil society organisations in peace-making and peacebuilding is often valorised, especially in countries that witness protracted conflicts" (Assal, 2016).

The role of civil society and community-based organisations in building resilience can be recognised to associate inter-linkages between DRR and human security protection, guided by the SFDRR "all-of-society approach" to DRR and development (Sendai Paragraph 7), and aligned with the 2018 High-Level Political Forum (HLPF) theme for Sustainable and Resilient Societies, as well as the (HLPF) 2019 theme for Inclusiveness and Equality. Noting that the High-level Political Forum is the United Nations central platform that was formed to follow-up and review the progress of member states of the 2030 Agenda for Sustainable Development and the Sustainable Development Goals (SDGs), it provides a platform for member state governments, UN system agencies, civil society, NGOs, and the private sector, to take stock of progress on the SDGs, discuss successes, challenges, and lessons learned in the process towards a fairer, more peaceful and prosperous world and a healthy planet by 2030.

Another aspect affecting the level and impact of the engagement of stakeholders in the assessment and building of resilience is the coordination and collaboration between different partners to remove silos, share resources, avoid duplication of efforts, and create new opportunities for learning and exchange knowledge. Understanding "what matters most" when building community resilience to disasters is key to identifying and critiquing how civil society and non-profit organisations are perceived by stakeholders and demonstrating how these organisations can enable communities to respond more effectively before (prevention), during (emergency humanitarian response), and after disasters (post-war recovery) (Roberts *et al.*, 2021). Further emphasis is placed on the roles and responsibilities of DRR key stakeholders in Sendai Paragraphs 36 and 48: "specific roles for individual stakeholder groups" in alignment with SDG 17, Target 17.17: Encourage and promote effective public, public-private and civil society partnerships, as well as by UNDRR representatives who are focused on filling the gap between national and local DRR platforms, building on the experience and resourcing strategies of partnerships and leadership provided to UN System partners at the global and regional levels:

> There is a lack of investments in local partnerships and bridging the gap between national and local DRR platforms, affecting the monitoring of resilience building at the local level, and the quality of data reporting on disaster data losses at the national level.
>
> (UNDRR, 2018)

In the post-disaster stage, there is a cross-cutting role of DRR key stakeholders and humanitarian aid agencies. Yet, depending on the scale of crisis and capacities of governmental institutions to respond to disaster, the level of humanitarian aid intervention can vary from one context to another and become more evident in fragile settings. In this situation, more agencies, such as the International Federation of Red Cross (IFRC), can be involved in providing basic needs for human security. Therefore, a strategic approach is required for managing mechanisms for coordination and collaboration between the different role-players to avoid overlapping of roles and responsibilities and prevent mismanagement of financial resources.

This is evident in the forming of the UN Cluster Approach, shown in Figure 7.3, as a standardised model for establishing such a mechanism. First formed following the 2005 earthquake in Pakistan, the clusters are groups of humanitarian organisations, both UN and non-UN, in each of the main sectors of humanitarian action, e.g., water, health, and logistics. They are designated by the Inter-Agency Standing Committee (IASC) and have clear responsibilities for coordination (OCHA, 2019). The United Nations Office for the Coordination of Humanitarian Affairs (OCHA) states that the aim of the cluster approach is

> to strengthen system-wide preparedness and technical capacity to respond to humanitarian emergencies and provide clear leadership and accountability in the main areas of humanitarian response. At the country level, it aims to strengthen partnerships and the predictability and accountability of international humanitarian action, by improving prioritisation and clearly defining the roles and responsibilities of humanitarian organisations.
>
> (OCHA, 2019)

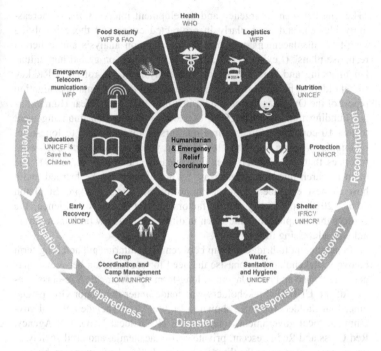

Figure 7.3 The UN Cluster Approach – Sectors of humanitarian action (OCHA, 2019)

Nevertheless, evidence from recent studies indicates that this approach does not provide guidelines on how to shift from humanitarian emergency response to long-term recovery and development and how aid finances are managed by the different stakeholders to build and maintain resilience for the most vulnerable population. Telford (2006) commented on the ability and will of international humanitarian agencies to contribute to disaster recovery and noted that "most humanitarian funds are donated for major emergencies, be they acute or chronic". This results in overlooking the impact of extensive, slow-onset disasters, and might generate highly irregular flows and levels of funding that can vary immensely from one emergency to another, especially if they occur far from donor borders. Geo-political interests of donors also have implications in defining the parameters of the emergency development nexus. For example, armed conflict was a major element of the crisis in the chronically under-funded emergencies in Darfur and the Democratic Republic of the Congo (Telford, 2006).

In the OECD 2017 Multi-Year Humanitarian Funding that was published as part of the World Humanitarian Summit, it was indicated that

"The line between emergency and development interventions is increasingly blurred, and particularly in protracted crises and those involving long-term displacement, which calls for common analysis and coherent response plans" (Levine *et al.*, 2019). With a focus on global humanitarian financing and how that affects the analysis of the role of DRR key stakeholders in building resilience, it was stated in the thematic evaluation report of the Overseas Development Institute 2019 Multi-Year Humanitarian Funding that "In 2016, 60% of global humanitarian financing went to just 10 countries. Almost three-quarters of this aid went to long-term recipients (where a crisis has lasted eight years or more) and 86% went to crises that had endured over three years" (Levine *et al.*, 2019). The main research question raised was: "What lessons can be learned about how to best enhance resilience in protracted crisis?" (Opitz-Stapleton *et al.*, 2019) to identify which factors shaped people's resilience, the choices that people can make when in difficulty, and how far humanitarian aid was addressing those factors.

In an attempt to bridge the gap between humanitarian aid and long-term recovery, stakeholders emphasise the need for coordination between emergency response and role-players in long-term development. A re-classification of the UNDRR Stakeholders was made under the Major Groups and other Stakeholders (MGoS), shown in Figure 7.2, to include national government, local government, international NGO, local NGO, UN Agency, Red Cross and Red Crescent, private sector, academia, and civil society as key sectors representing the three units of study (local governments, humanitarian aid, and IDPs and refugees) to investigate the level of responses and participation of DRR stakeholders better and avoid the overlap between the UN humanitarian agencies and local role-players in the processes assessing and building resilience (Helbig, 2015). The re-classification process of DRR key stakeholders was followed with a mapping exercise to link the stakeholders identified and their roles in achieving each of the Sendai four priorities of action: Priority 1: Understanding disaster risk; Priority 2: Strengthening disaster risk governance to manage disaster risk; Priority 3: Investing in disaster risk reduction for resilience; and Priority 4: Enhancing disaster preparedness for effective response and to "Build Back Better", as shown in Figure 7.4.

Another issue relevant to debate about the coordination and collaboration of DRR key stakeholders is the conflict of powers.

> Conflicts arise when the decisions are opposite to the interests of the stakeholders. Stakeholders with higher political or institutional hierarchy may attempt to use their power and political influence, to discredit the decision and eventually change it according to their interests.
>
> (Moura & Teixeira, 2009)

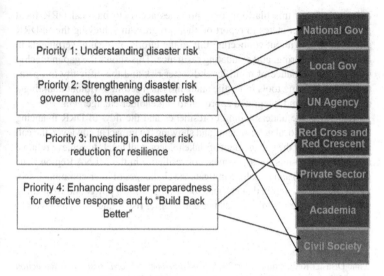

Figure 7.4 Connectivity between DRR stakeholder and the SFDRR four priorities of action (UNDRR, 2015)

These theoretical insights were also reflected from field-based observations by the Yemeni Red Crescent representative, from a fragile and violent conflict background, as stated:

> Community leaders who participate on behalf of their people can sometimes redirect the real community views, as they absorb the proposed humanitarian projects at a certain level, and observe benefits from a certain angle, without allowing you to reach the real level of vulnerability 'blocking access'. Community leaders in fragile contexts can become an obstacle for development – Power of control.
>
> (Yemen Red Crescent, 2018)

Following the sectoral analysis of DRR key stakeholder groups, it is critical to apply a geographical classification of DRR key stakeholders at local, national, and regional levels in order to understand the scale of representation, the overlap of functions between different sectors, their level of involvement in DRR stages, and how this can feed into the process of assessing resilience and the decision-making process for building resilience. Another issue relevant to this debate is the lack of engagement of DRR stakeholders at national level, as the process of assessing resilience is applied at the local level (Dalal-Clayton, 2002). Despite launching the UNDRR SENDAI Monitoring Framework online platform, this platform only provides access to national DRR focal points of member states to report on their progress in achieving the SFDRR global targets, without connecting this to the efforts of local governments in building, monitoring, and evaluating resilience. This shows the gap in mechanisms for governance of multi-level disaster risk and the difficulty in operationalising existing tools in fragile and conflictual settings without adopting a more inclusive and integrated approach to stakeholder engagement.

Regarding the understanding of resilience and the flow of DRR financing from national to local governments and showing the impact at city level, the call in this book is for the engagement of stakeholders at all levels to achieve reliable and impactful results for assessments of resilience in the MENA Region (Virginia *et al.*, 2017) and to help to mitigate the challenges of inter-communal, sectarian, and tribal conflict that can lead to shifting political power and weakening urban governance, building resilience and disaster risk reduction at all levels.

Reference list

Asian Disaster Reduction Center. 2015. *Sendai Framework for Disaster Risk Reduction 2015–2030*. United Nations Office for Disaster Risk Reduction.

Assal, M. A. M. 2016. Civil society and peace building in Sudan: A critical look. *Sudan Working Paper*. Available Online: https://www.cmi.no/publications/file/5807-civil-society-and-peace-building-in-sudan.pdf

Booth, L., Schueller, L. A., Scolobig, A., & Marx, S. 2020. Stakeholder solutions for building interdisciplinary and international synergies between climate change adaptation and disaster risk reduction. *International Journal of Disaster Risk Reduction*, 46: 101616.

Brugha, R., & Varvasovszky, Z. 2000. Stakeholder analysis: A review. *Health Policy and Planning*, 15(3): 239–246.

Chevalier, J. M. 2002. *Natural Resource Project/Conflict Management: Stakeholders Doing "Class Analysis"*. Available Online: https://www.academia.edu/73647093/Natural_resource_project_conflict_management_stakeholders_doing_class_analysis

Cleland, D. I. 1997. Project stakeholder management. *Project management handbook*, 275–301.

Dalal-Clayton, D. B., Swiderska, K., & Bass, S. (Eds.). 2002. *Stakeholder Dialogues on Sustainable Development Strategies: Lessons, Opportunities and Developing Country Case Studies*. International Institute for Environment and Development.

Expert from the Overseas Development Institute. 2017. (Interview, 23 June).

Expert from the Yemen Red Crescent. 2018. (Interview, 04 March).

Fraser, A., & Kirbyshire, A. 2017. *Supporting Governance for Climate Resilience: Working with Political Institutions*. Available Online: https://odi.org/en/publications/supporting-governance-for-climate-resilience-working-with-political-institutions/

Freeman, R. E. 1984. *Strategic management: A stockholder approach*. Pitman.

Freeman, R., Edward, J., Harrison, S., Wicks, A. C., Parmar, B. L., & De Colle, S. 2010. *Stakeholder Theory: The State of the Art*. Available Online: https://core.ac.uk/download/pdf/346447581.pdf

Helbig, N., Dawes, S., Dzhusupova, Z., Klievink, B., & Mkude, C. G. 2015. Stakeholder engagement in policy development: Observations and lessons from international experience. In: *Policy Practice and Digital Science: Integrating Complex*

Systems, Social Simulation and Public Administration in Policy Research, pp. 177–204. Springer International Publishing.

Lafrenière, J., Sweetman, C., & Thylin, T. 2019. Introduction: Gender, humanitarian action and crisis response. *Gender & Development*, 27(2): 187–201.

Levine, S., Sida, L., Gray, B., & Venton, C. C. 2019. *Multi-Year Humanitarian Funding: A Thematic Evaluation*. Available Online: https://odi.org/en/publications/multi-year-humanitarian-funding-a-thematic-evaluation/

Littau, P., Jujagiri, N. J., & Adlbrecht, G. 2010. 25 years of stakeholder theory in project management literature (1984–2009). *Project Management Journal*, 41(4): 17–29.

Moura, H. M. and Teixeira, J. C. 2009. Managing stakeholder conflicts. *Construction Stakeholder Management*: 286–316.

Murray, V., Maini, R., Clarke, L., & Eltinay, N. 2017. Coherence between the Sendai framework, the SDGs, the climate agreement, New Urban Agenda and World Humanitarian Summit, and the role of science in their implementation. *Proceedings of the Global Platform for Disaster Risk Reduction, Cancun, Mexico*, pp. 24–26. Available Online: https://www.preventionweb.net/publication/coherence-between-sendai-framework-sdgs-climate-agreement-new-urban-agenda-and-world

Opitz-Stapleton, S., Nadin, R., Kellett, J., Calderone, M., Quevedo, A., Peters, K., & Mayhew, L. 2019. *Risk-informed development: from crisis to resilience*. ODI/UNDP, New York.

Ramirez, R. 1999. Stakeholder analysis and conflict management. In: *Cultivating Peace: Conflict and Collaboration in Natural Resource Management*. IDRC.

Roberts, F., Archer, F., & Spencer, C. 2021. The potential role of non-profit organisations in building community resilience to disasters in the context of Victoria, Australia. *International Journal of Disaster Risk Reduction*, 65: 102530.

Stein, S., & Walch, C. 2017. The Sendai framework for disaster risk reduction as a tool for conflict prevention. In: *Conflict Prevention and Peace Forum*. Available Online: https://www.preventionweb.net/publication/sendai-framework-disaster-risk-reduction-tool-conflict-prevention

Swiderska, K. 2002. *Implementing the Rio Conventions: implications for the south*. International Institute for Environment and Development.

Telford, J. 2012. *Disaster Recovery: An International Humanitarian Challenge?* Available Online: https://doi.org/10.1355/9789814345200-007

Telford, J., & Cosgrave, J. 2006. Joint evaluation of the international response to the Indian Ocean tsunami: synthesis report. Tsunami Evaluation Coalition (TEC).

Trogrlić, R. Š., Cumiskey, L., Triyanti, A., Duncan, M. J., Eltinay, N., Hogeboom, R. J., Jasuja, M., Meechaiya, C., Pickering, C. J., & Murray, V. 2017. Science and technology networks: A helping hand to boost implementation of the Sendai Framework for Disaster Risk Reduction 2015–2030?. *International Journal of Disaster Risk Science*, 8: 100–105.

United Nations Office for the Coordination of Humanitarian Affairs (OCHA). 2019. *The Cluster Approach*. Available at: www.humanitarianresponse.info/en/about-clusters/what-is-the-cluster-approach

United Nations Office for Disaster Risk Reduction (UNDRR). 2015. *Sendai framework for disaster risk reduction 2015–2030*. United Nations Office for Disaster Risk Reduction: Geneva, Switzerland.

United Nations Office for Disaster Risk Reduction (UNDRR). 2018. *Stakeholder Advisory Group*. Available at: www.unisdr.org/archive/57204

United Nations Office for Disaster Risk Reduction. 2019. *Partners and Stakeholders*. Available at: www.undrr.org/implementing-sendai-framework/partners-and-stakeholders

8 Urban Resilience – Opportunities and Constraints for Adaptation to Climate Change

Monitoring Protracted Displacement

Evidence from the Migration, Environment and Climate Change Report of the International Organisation for Migration (IOM, 2009) indicated:

> a call for better data to answer questions relating to the likely scale and pattern of movement such as, how many will migrate due to environmental/ climate change? Who will migrate? When and where will they migrate; will new destinations have to be found? Will migration be temporary or permanent, internal or international? What will be the consequences of migration for the people who move, for those left behind and for the places of destination? There is also a concern to understand better the here and now – how is environmental change affecting migration today and can we already identify especially vulnerable populations or regions?
>
> (IOM, 2009)

When displacement is first triggered by violent conflict, IDPs and refugees are hosted in camps by emergency aid agencies to provide refuge from violence, access to food, services, and temporary shelter. Once conflict comes to an end, displaced people tend to return to their places of origin as one of the first options explored for a "durable solution". This action takes place as the first response to secure ownership, revive previously existing economies, reclaim land rights, and re-establish sustainable livelihoods. Nevertheless, in fragile states, returning IDPs and refugees are faced with different scenarios, including the realities of destruction of infrastructure and the loss of livelihoods on top of crushing poverty, landmines, property restitution issues, incomplete disarmament processes, and political obstacles. The resulting combination turns re-integration and resettlement into a very fragile and years-long process, laying the ground for a prolonged displacement cycle known as "protracted displacement". Thus, in this chapter, examples of protracted displacement in the MENA Region are discussed, as well as the need for framing durable solutions for climate security displaced people.

DOI: 10.4324/9781003363224-8

The definition of "protracted displacement" was agreed upon by participants at the 2007 expert seminar on protracted IDP situations, hosted by the UNHCR and the Brookings-Bern Project on Internal Displacement: "Protracted internal displacement situations are those in which the processes of finding durable solutions have stalled and/or IDPs are marginalised because of violations or a lack of protection of human rights, including economic, social and cultural rights". Data gaps in monitoring protracted displacement and socio-economic determinants of vulnerability were also investigated while exploring the tools and methodologies applied by humanitarian aid and international development agencies to monitor IDPs during disasters in urban settings (Benson *et al.*, 2007).

In terms of internally displaced persons (IDPs) "who are forced to flee their homes due to armed conflict, generalised violence, violations of human rights, natural or human-made disasters, but who remain within their own country" (OCHA, 2018), the total number of IDPs has doubled over the past 15 years from below 20 million in the 1990s to 27.5 million by 2010. Five years later, internal displacement caused by conflict and violence has reached more than 40 million, particularly in the Middle East following the Arab Spring uprisings that began in late 2010. Having considered conflict and violence, displacements caused by disasters also increased, as stated in the 2014 Global Report of the Internal Displacement Monitoring Centre: "such events triggered the displacement of 20.7 million people, or 94 per cent of the global total" (IDMC-Disasters, 2014). Figures from the 2016 report also accentuated that "there were 19.2 million new displacements associated with disasters brought on by rapid-onset natural hazards in 2015, more than twice as many as for conflict and violence" (IDMC, 2016).

Using the IDMC reports as the primary sources of data collection, it is noted that there was no such evidence base for internal displacement at the time, but the first Global IDP Survey was undertaken in 1997–98. This gave rise to the Global IDP Project in 1998, which later became the Internal Displacement Monitoring Centre (IDMC). Since then, several annual global figures and analyses of patterns and trends for internal displacement associated with conflict were published until the year 2008, when monitoring of displacement caused by disasters began. "Even today, however, there are major evidence gaps on local dynamics and global trends, there are numerous challenges in collecting and analysing basic metrics such as the number of IDPs, their locations and the duration of their displacement" (Off the Grid, IDMC, 2018). An important contribution offered in this chapter is the statement provided by a representative from the IDMC, who is the coordinator of the process and production of the 2018 Report/The Grid, about the validity of the data:

What kind of makes us different from IOM or UNHCR, or even UNDRR, when it comes to disasters, is what we call data triangulation. We cross-check data. If you had a look at our work, you will see a regional breakdown and a country portfolio, so it's basically what we know. But the 2018

GRID report has a full chapter about what we don't know, and we high-lighted very clearly, Yemen. Last year, we saw urban warfare as maybe we've never seen since World War Two. They've been classified as a level three emergency by the United Nations, and it's very difficult for our part-ners involved in data collection to get actual data on what is going on in terms of the people displaced. Mostly in Sana in Yemen not only the newly displaced, but those who are secondary displaced. And even in Aleppo and Eastern Ghouta in Syria, we had a series of urban conflicts that resulted in large-scale displacement. Apart from those that benefited from humanitar-ian assistance in mainly IDP camps, it was extremely difficult for us to provide, obtain data or provide a precise picture about those who were not in camps. It is basically because people are settling informally in the outskirts of the city and staying with host families. There's a lot of intra-city movement, and this is why we would like to open the discussion about urban internal displacement.

(IDMC, 2017)

Factors regarding the amount of time in protracted displacement and the number of people affected in urban settings remain missing from data about displacement and must be considered in determining whether a situation is protracted (IDMC, 2008). Monitoring data about protracted displacement will pursue the utilisation of existing tools assessing resilience, such as the Disas-ter Resilience Scorecard – New Ten Essentials of the United Nations Office for Disaster Risk Reduction (UNDRR, 2017), to provide better understanding of underlying drivers of displacement risk, help to build resilient urban devel-opment, and provide sustainable access to land ownership, critical infrastruc-ture, and security of tenure. This approach demonstrates the inter-relationship between the indicators covering the needs of IDPs and potential variables required to address sustainability challenges: data structuring (informal set-tlements), operationalisation (land-use planning), accountability (land tenure security), interpretation (property rights), and gaps in data about losses cause by disaster (displacement).

Protracted Displacement in the MENA Region

To investigate the phenomenon of protracted displacement and transformation of camps from temporary shelters into permanent settlements, Lebanon was selected as a case study, having the fourth-highest per capita concentration of refugees in the world, with 1 million according to the UNHCR figures for June 2018 (UNHCR, UNICEF & WFP, 2018). In this case, the relationship between hazards, exposure, and vulnerability to disaster risk and violent conflict was investigated through the socio-spatial theories and key debates about "exigent cities" and "emergency urbanism" (Sanyal, 2017). The case study of Lebanon helped to develop an evidence-based policy guideline for U-RAP on how best to pursue disaster risk reduction in fragile contexts, challenged with the complexity of refugee camps, sovereignty, and informality beyond the jurisdiction system of the state. In Lebanon, during the first post-independence presidential terms (1943–1952 and 1952–1958) significant demographic changes in the population of Lebanon were experienced with the influx of Armenian and Syrian Political Orthodox Christian refugees who fled Turkish persecution, followed by the wave of 100,000 Palestinian refugees who fled the first 1948 Arab–Israeli war. These numbers are in rise with the recent Israel and Palestine Conflict escalation in Oct 2023.

With the continuing influx of refugees, clear class and sectarian discrepancies between different community groups grew, triggering the Lebanese civil war in April 1975, which was a culmination of heightened tension between the Lebanese Christian side and the Muslim-Leftist side and their Palestinian allies. The Palestinian refugee camps, initially formed as recruitment and training centres for Palestinian commando fighters to be part of the battle against Israel, acted as hubs for attracting

> the Muslim Lebanese youth in the poverty-stricken quarters neighbouring these camps, who found in the Palestinian struggle a space of rebellion against traditional and skewed politics and injustices caused by the Lebanese State's neglect. For Christians, on the other hand, the militarization of Palestinian factions and their Muslim allies, and their visibility in streets and neighbourhoods was seen as a tangible threat to a state they had sought to build.
>
> (Yassin, 2012)

During the same period, the Israeli invasion in 1982 and Syrian armed control of the Lebanon western side, as of 1986, left more than 150,000 killed, 300,000 injured, and approximately 800,000 displaced. The civil war lasted for two years and generated new urban space and territories according to the emerging political, sectarian, and military realities (Yassin, 2012), with divisions taking place until a national reconciliation peace settlement was reached in 1990 under a modified power-sharing mechanism based on equal Muslim and Christian representation.

Identifying links between the context of protracted displacement in these countries and building urban resilience, it is noted that Syria was one of the first countries in the MENA Region to take the lead in implementing DRR activities following the launch of the MCR Campaign in Kuwait with the host of the Arab Towns Organisation and participation of representatives of Arab cities from Aleppo (Syria), Amman (Jordan), and Tripoli (Lebanon) on 3 October 2010. Ten days later, Aleppo celebrated the International Day for Disaster Risk Reduction, and launched the 'Safety in Prevention' poster through the Child Friendly Schools Programme of United Nations Children's Fund (UNICEF), with the aim of applying DRR activities to raise awareness of school children through evacuation drills in collaboration with the civil defence, firefighting department and ambulance services (UNDRR, 2010).

Further engagements with global initiatives to build resilience, such as the City Profiles Programme of UN-Habitat, were also traced in the same countries. "UN-Habitat City Profiles are formulated to offer a cross-sectoral perspective on urban vulnerabilities that will inform holistic and inclusive interventions by local authorities, humanitarian partners and others to respond to needs and alleviate poverty amongst host and displaced populations" (UN-Habitat, 2016). In the MENA Region, the profiles of Arab cities completed during the period 2014–2018 were: Homs (Syria, 2014), Aleppo (Syria, 2014), Mosul (Iraq, 2016), Tripoli (Lebanon, 2017), Tyre (Lebanon, 2017), Maachouk Neighbourhood, Tyre (Lebanon, 2017), Nabaa Neighbourhood, Bourj Hammoud (Lebanon, 2017), Tabbaneh Neighbourhood, Tripoli (Lebanon, 2018), El-Qobbeh Neighbourhood, Tripoli (Lebanon, 2018), and Jabal Mohsen Neighbourhood, Tripoli (Lebanon, 2018). It is noted that seven out of the ten cities are based in Lebanon, which shows the unbalanced distribution of the implementation of the initiative across the MENA Region. On the other hand, the common characteristic between all of the cities is the fragile context of post-war and the challenge of the refugee crisis, indicating that without focusing on the root problems of protracted displacement, principles of resilience cannot be translated into action.

Not forgetting the impact of the Syrian crisis on a regional scale, views from the Syrian Red Crescent were captured to understand the scale and scope of building resilience in such a fragile setting. Yet, it is worth noting the variations in governance of the Arab states of the Syrian refugees' crisis with a contrasting camp policy in country borders with Syria (e.g., Jordan) and non-camp policy (e.g., Lebanon). With regard to the Lebanese approach of keeping the borders open for Syrian Refugees until October 2014, it is important to highlight the great impact on the stability of the country and the complexity of the conditions of pre-existing Palestinian refugees in Lebanon, as well as how this affects vulnerabilities of refugees to disaster and their conditions in both camp and non-camp settings (informal gatherings) in Lebanon.

More to the point is that the management of the Syrian crisis in the Middle East is very much informed by previous experiences with refugees in countries such as Jordan and Lebanon, particularly the prolonged 70 years of the Palestinian situation. With both countries being non-signatories of the 1951 Refugee Convention, limiting the rights of Palestinian refugees for land ownership and access to employment was embraced strategically to support the refugees' demand of the "right to return". Yet, this flexibility of crafting their rights created grey spaces of legitimacy and lack of legal status and protection in the relationship between the state, citizens, and refugees, creating marginalised communities that are greatly exposed to the risk of urban violence and vulnerability to disasters (Baytiyeh, 2017).

Lebanon's pre-existent socio-economic challenges, characterised by low economic activity rates, high youth unemployment, and a large informal economy, were all aggravated by the Syrian crisis. It had a profound impact on poverty and inequality, and unemployment among Palestinian refugees in Lebanon rose to 23% in 2015, compared with 8% at the start of the Syrian crisis. This was because of the willingness of the new arrivals to work for less and under more unfavourable conditions. These factors also resulted in an increase in poverty rates. Currently, 90% of Palestinian refugees from Syria and 68% of pre-existing Palestinian refugees live below Lebanon's poverty line of 3.84 US Dollars per day, while 6% of Palestinian refugees from Syria in Lebanon live in extreme poverty.

All the factors above caused the intensification of competition over infrastructural services, jobs, and accommodation, disproportionately affecting the most vulnerable and jeopardising the access of refugees to humanitarian aid while increasing the pressure and visibility of urban informality, exposing the most vulnerable communities to the disaster risk of natural and man-made hazards. The points outlined above were reflected in the views of an expert from ODI, who worked in the MENA Region and shared experiences of shifting from dependence on humanitarian aid to development and the role of aid agencies:

> One of the basic elements is that humanitarian interventions need to be politically neutral. If you're operating in a conflict zone or humanitarian earthquake, you just deliver to whoever needs, which varies from developmental system. The point about humanitarian intervention is to just give them what they need. They need food the first thing, give it to them. Now, with development, the approach is different in an important way, which is you're trying to think about long-term sustainability. So you don't just give people things. You try to create an institutional setting which enables long-term sustainability. So, if you give something to somebody, they may expect that then to be free in the future. So, you're very wary about creating this kind of sense of dependency or entitlement.
>
> (ODI, 2018)

Property Rights – Land Tenure

An overview of the international conventions for people who lose access and rights to land following a disaster will apply in the context of IDPs. In the Guiding Principles on Internal Displacement – Principle 6 – Section 1, it is indicated that: "Every human being shall have the right to be protected against being arbitrarily displaced from his or her home or place of habitual residence" (UNHRC, 1998). Nevertheless, the lack of legal documents and ownership of land for displaced people cause the difficulty to apply the rule of "his or her home or place" and can generate new challenges to achieving durable solutions for protracted displacement.

This applies to Principle 21, in which it is stated that "property and possessions left behind by internally displaced persons shall be protected against destruction and arbitrary and illegal appropriation, occupation or use". For the refugees, in the 1951 Convention related to the status of refugees, a refugee was defined as a person who has crossed international borders as a result of a "well-founded fear of being persecuted" on account of their religious, political, sexual, or other social identity, and whose country will not or cannot protect them or might even be the body that is persecuting them (United Nations General Assembly, 1951, Hynie, 2018). Thus, in this chapter, the challenges and opportunities of IDPs and refugees are explored in the context of displacement caused by disaster while exploring their rights and needs to build urban resilience.

Nevertheless, the Universal Declaration of Human Rights (United Nations, 1948) provided another perspective under Article 17 that the people affected by a disaster have the right not to be deprived arbitrarily of their property. This focus on the institutional governance of the city and its role in providing the rights for displaced people according to the Pinheiro Principles means that displaced people have the right to return to their lands when the emergency response is completed. Each person should have rights to land that are at least as good as the situation prior to the disaster. In Principles 28–30 of the UN Guiding Principles on Internal Displacement, it is stated that "competent authorities have the responsibility to assist displaced persons to recover their property or, where this is not possible, to assist these people in gaining appropriate compensation or just reparation. The land policy framework should be consistent with these principles" (Mitchell, 2011).

In terms of identifying the challenges for Climate Security Displaced (CSD) people, vulnerabilities and capacities can be assessed using different tools and indices but, as previously indicated in the literature review (Chapter 3), "researchers found that there are frequent gaps and incoherencies between the asserted definitional and contextual meanings of resilience/vulnerability and their implementation – particularly the absence of explicit frameworks" (Ionescu et al., 2009).

There is a wide range of tools available, "but often in practice so much information is collected, on so many different issues, and of such diversity (including variations in quality), that it is difficult to shape a coherent analysis" (Benson et al., 2009; Twigg, 2009). In 128 instances of assessments of vulnerability, Zou and Thomalla (2008) found only 14% referring to a framework for vulnerability (Gall, 2013). Thus, the scope of this study moved beyond identifying the rights and needs using assessments of vulnerability, towards understanding the variations in challenges and vulnerabilities between different contexts of IDPs and refugees in the MENA Region and incorporating those components into policies and action plans for resilience. Evidence from an ODI representative supports the research decision as below:

> My understanding is the humanitarian system has got a pretty well-defined set of assessment criteria, and service delivery criteria for basic services. There is now an honourable tradition, but there are problems with it because you really

got to understand the dynamics of the communities, the institutions within the community and the institutions within which the community exists.

(ODI, 2018)

Another issue relevant to this debate is the analysis of data about vulnerability and capacity and the transfer of results from assessments of vulnerability to effective decision-making frameworks, which must be sustained with understanding of the local context, and associate links with Urban Resilience Action Plans (U-RAP) that can capture the dynamics of displacement of IDPs and refugees caused by disasters in urban settings. These theoretical underpinnings were supported by the practical experience of the IFRC representative who applied VCAs in disaster and fragile conflict settings:

> We usually do risk assessment using the Red Cross tool (VCA) Vulnerability and capacity assessment framework. In a conflict setting, you have to be very careful when you discuss certain issues with the people, because it's an open community meeting where different ethnic groups or conflicting parties are also there, you need to have specific facilitation skills to understand what divides people and what connects them.

(IFRC, 2018)

Shaped by the type of hazards and degree of exposure, it is useful to consider the combination between social and bio-physical vulnerabilities in assessing resilience by assessing the social structure and demographic profile of IDPs and refugees inhabiting informal settlements to help to understand the relationship between vulnerability and resilience in fragile and conflict settings (Beatley, 2014). "The lack of secure property rights severely hinders the displaced people's ability to use land and real estate as collateral to access finance" for long-term recovery and development of livelihood (Madbouly, 2009).

Land tenure plays a vital role in building resilience and understanding of risk in the local context of the MENA Region. As indicated by Adoko and Levine (2004), "displacement continues beyond a short-term period, it is necessary to create measures to protect the people's rights to their land so that they have a level of security while waiting for restitution in the internally displaced people (IDPs) camps" (cited in Mitchell, 2011).

Defined by the Food and Agriculture Organisation (FAO), land tenure is "the legal or customary relationship among people with respect to land and associated natural resources such as water, trees, minerals or wildlife". Ruling how property rights are allocated and access to resources is regulated, land tenure is used to normalise the relationship between people and land in the event of disasters and to provide the power structures for managing resources within a society (Mitchell, 2011). In the conditions of informal settlements and emergency shelters for displaced people, the challenges of land ownership increase, thus the property rights and building legislations for displaced people in pre-and-post disaster settings are considered extensively in the context of the MENA Region (Arab States). In this region, complex, customary land rights dominate land tenure security, associated with the dispute resolution systems for the deprived and the lack of legal recognition for people displaced by disaster.

Land use planning refers to:

> The process undertaken by public authorities to identify, evaluate and decide on different options for the use of land, including consideration of long-term economic, social and environmental objectives and the implications for different communities and interest groups, and the subsequent formulation and promulgation of plans that describe the permitted or acceptable uses.
>
> (Mitchell, 2011)

From that perspective, the indicators for Essential 4: Pursue Resilient Urban Development, were investigated in relation to the integration of previously outlined themes of displacement, informal settlements, land-use planning, land tenure security, and property rights, as shown in Table 8.1.

Table 8.1 Sustainability Assessment for UNDRR Scorecard Essential 4 – Indicators

Essential 4: Pursue Resilient Urban Development

Evaluation: Displacement – Informal Settlements – Land-use Planning – Land Tenure Security – Property Rights

Indicators	Question/Assessment Area	Considerations	Evaluation
Land-use Zoning	Is the city appropriately zoned, considering, for example, the impact from key risk scenarios on economic activity, agricultural production, and population centres?	Displacement for three months or longer as a consequence of housing being destroyed or rendered uninhabitable or the area in which it is located being rendered uninhabitable. This assessment must also cover informal and unplanned settlements.	Displacement Informal settlements
New Urban Development	Are approaches promoted through the design and development of new urban development to promote resilience?	Is there a policy promoting physical measures in new development that can enhance resilience to one or multiple hazards? For example, appropriate locations for new development, water-sensitive urban design, proper integration of disaster refuge areas, proper access and egress routes.	Land-use planning

(Continued)

Table 8.1 (Continued)

Essential 4: Pursue Resilient Urban Development

Evaluation: Displacement – Informal Settlements – Land-use Planning – Land Tenure Security – Property Rights

---

Indicators	Question/Assessment Area	Considerations	Evaluation
Building Codes and Standards	Do building codes or standards exist and do they address specific known hazards and risks for the city? Are these standards regularly updated?	Standards will include those for the supply of basic infrastructure services to informal settlements, without which the ability of those settlements to recover from disasters will be severely compromised.	Land Tenure security
Application of Zoning, Building Codes, and Standards	Are zoning rules, building codes, and standards widely applied, properly enforced, and verified?	Cities with informal settlements are unlikely to score highly on this assessment unless the occupants of those settlements have been engaged and helped in making themselves more resilient.	Property Rights

---

An important contribution offered in this chapter is the application of the principles of sustainability assessment to the SFDRR Disaster Resilience Scorecard. In association with land-use rights, this criterion addresses the long-term sustainable development needs for the most vulnerable people, "taking into consideration how low-income groups can access suitable land" (UNDRR, 2017). The selection of the terminology in Table 8.1 is framed according to the five purposive challenges of sustainability assessment determined by Waas *et al.* (2014) for "the decision-making strategy in sustainable development as follows: information structuring – operationalisation – accountability – data gaps".

Under Essential 4, it is important to consider that the vulnerability and exposure of IDPs and refugees living in "informal settlements" to disaster vary considerably from those living in "slums", which are commonly linked. Madbouly (2009) states "informal" as being "the housing stock which is not in compliance with current regulations", where "slums" refer to "deteriorated living conditions and low levels of access to basic services" (Madbouly, 2009).

The understanding of definitions and taxonomies of terminology in the fragile context of the MENA Region is strongly emphasised in this study to analyse the characteristics and normative nature of the rights of displaced people and to investigate how risk-aware urban planning, design, and implementation for sustainable development are essential for "addressing the needs of informal settlements, including basic infrastructure deficits such as water, drainage and sanitation" (UNDRR, 2017). Accordingly, the guidance of the Food and Agriculture Organisation (FAO) was used to define "land tenure security" and "property rights". Land tenure security is defined as "the certainty that an individual's rights to land were recognized by others and protected in cases of specific challenges", while "property rights" are defined as "recognized interests in land or property vested in an individual or group" including "customary, statutory or informal social practices which enjoy social legitimacy at a given time and place" (Mitchell, 2011).

This approach demonstrated the inter-relationship between the indicators that address the needs of IDPs, and potential variables required to address the challenges of sustainability: data structuring (informal settlements), operationalisation (land-use planning), accountability (land tenure security), interpretation (property rights), and gaps in losses of data about disaster (displacement). The first challenge is data structuring, where socio-economic vulnerability of informal settlements is difficult to measure, verify, and communicate as meaningful information for the decision-making process. The second challenge is the operationalisation of the indicators, where land-use planning pulls "the discussion of sustainable development away from disaster emergency response abstract formulations" (Waas *et al.*, 2014).

Changing the mindset of decision-makers is a priority for challenge three, to ratify the rights of displaced people for land tenure security. Challenge four demonstrates the accountability for property rights and leads to the fifth challenge of data gaps about losses caused by disaster for the first evaluation variable of displacement. Ratifying agreements signed internationally and "hard laws" will help to secure land and property rights for displaced people as part of this assessment. Nevertheless, the role of "soft laws" and legal frameworks for disaster risk reduction should not be ignored. The monitoring of progress against the 2015–2030 global indicators helps to raise awareness and understanding of existing challenges and opportunities while developing a culture of social learning, as argued by Pintér *et al.*, "the way society measures progress represents a key leverage point in tackling the root causes of unsustainable development" (cited in Waas *et al.*, 2014).

Displacement Durable Solutions

The review and analysis of the laws and mechanisms to integrate the vulnerabilities of IDPs and refugees into interventions in policy for disaster risk reduction to achieve urban resilience in this chapter guides the investigation of the opportunities and challenges facing IDPs and refugees. The interconnections between climate change, conflict, and displacement integrate the knowledge gained from the collection and analysis of primary and secondary data for building urban resilience. With a focus on building societal capacity and filling the gaps in existing indicators used to assess resilience, the use of open data is examined to validate the input and output of the assessments of urban resilience and to define the parameters for DRR and Climate Change Adaptation (CCA) Policy by applying a correlation between the indicators for the Sendai Monitoring Tool, Disaster Resilience Scorecard for Cities, and durable solution principles for Climate Security Displaced (CSD) people.

An overview of the international conventions for people who lose access and rights to land following a disaster is provided in the context of Climate Security Displaced (CSD) refugees and IDPs. The findings of the literature review about the definition and dynamics of resilience discussed in Chapter 3 is included, together with risk and vulnerability to highlight the needs of IDPs and refugees affected by displacement caused by both natural disasters and conflict. Bearing in mind the disaster emergency response of international humanitarian aid agencies, protection efforts are directed mostly towards supporting the temporary settlement of refugees and voluntary return of IDPs. Yet, the impact of disaster risk management and protracted displacement on the scale of city and urban spatial planning is often ignored, with social integration and access to infrastructural services outlined only from the broader perspective of camps and shelters, overlooking the impact of disaster risk reduction on human rights for access to land and security of tenure.

In the IASC Framework on Durable Solutions for Internally Displaced Persons, it is recognised that "a durable solution is achieved when internally displaced persons no longer have any specific assistance and protection needs that are linked to their displacement, and can enjoy their human rights without discrimination on account of their displacement" (The Brookings Institution – University of Bern Project on Internal Displacement, 2010). Further investigation into how this concept is interpreted in real contexts is applied by capturing the qualitative views of representatives from international and local agencies involved in researching and supporting IDPs and refugees to achieve durable solutions.

It's important to open a discussion on the present internal displacement jargon of "durable solutions", and the complexities that go beyond counting. There are three types of durable solutions, it can be local integration, relocation, or return. For the Middle Eastern context, what do we mean by durable solutions, because it seems that local integration is being the *de facto* for durable solution, and is it durable? It is not only about the number of people, but it's about the conditions ultimately these people live in. We are even entering into the discussion of system dynamics modelling, when do you move?

(IDMC, 2018)

The contributions of the International Organisation of Migration (IOM) towards the progressive resolution of displacement situations were mapped against the eight criteria outlined in the Inter-Agency Standing Committee (IASC) Framework on Durable Solutions for Internally Displaced Persons (Figure 8.1).

Kivelä *et al.* (2018), presented the IASC Analytical Framework for Durable Solutions Analysis as part of the publication of the Durable Solutions Analysis Guide. Written based on learning from the project informing responses to support durable solutions for IDPs implemented by the Joint IDP Profiling Service, the guide provided evidence on the perspectives of displaced people on durable solutions, including which settlement option to pursue and how this is aligned with the eight criteria of the IASC framework that determine the extent to which a durable solution has been achieved. Based on the core demographic data of the displaced population, the mandate of this guide is to support a special rapporteur on the human rights of Internally Displaced Persons. Insights into understanding the meaning of "durable solutions" from the Internal Displacement Monitoring Centre (IDMC) (Figure 8.2), associated with the principles of assessing sustainability, guided the investigation in this study of the factors and drivers affecting the decision-making process for durable solutions in urban settings.

An important contribution to the body of knowledge offered in this publication is demonstrated in applying the principles of "sustainability assessment" to assessments of urban resilience (Chapter 4) using the SFDRR Disaster Resilience Scorecard Indicators. Thus, the environmental, social, and economic complexities of the hazards of disaster risk, exposure, and vulnerability cannot be captured without the wider interpretation of sustainability in the decision-making process. Nevertheless, further study worthy of consideration is the integration of qualitative methods, beyond the quantitative indicators of the SFDRR and SDGs, to measure the human experiences of DRR key stakeholders (Chapter 7) in the process of decision-making about resilience and to provide evidence based on the negative impact of adopting exclusive citizen engagement in the ownership of executing the practices of action plans for resilience.

The contextualisation of mechanisms for data collection was recommended to maximise the efficacy and productivity of using "vulnerability and capacity risk assessments" in order to capture the changes in the needs of IDPs and refugees across time and space. Integrating the principles of "safer access" will help to engage informal role-players in gathering evidence-based data that reflect the real context of vulnerable communities. Strengthening institutional capacity for resilience is the core of building resilience of IDPs

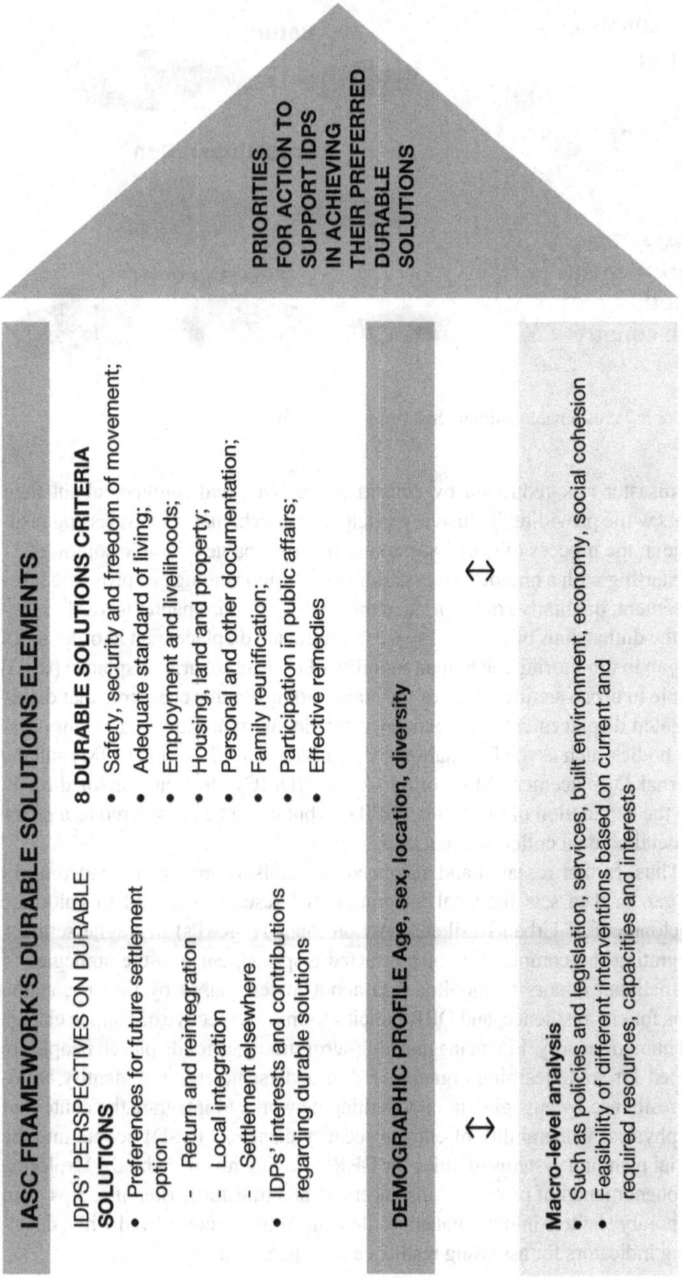

IASC FRAMEWORK'S DURABLE SOLUTIONS ELEMENTS

IDPS' PERSPECTIVES ON DURABLE SOLUTIONS

- Preferences for future settlement option
  - Return and reintegration
  - Local integration
  - Settlement elsewhere
- IDPs' interests and contributions regarding durable solutions

8 DURABLE SOLUTIONS CRITERIA

- Safety, security and freedom of movement;
- Adequate standard of living;
- Employment and livelihoods;
- Housing, land and property;
- Personal and other documentation;
- Family reunification;
- Participation in public affairs;
- Effective remedies

PRIORITIES FOR ACTION TO SUPPORT IDPS IN ACHIEVING THEIR PREFERRED DURABLE SOLUTIONS

DEMOGRAPHIC PROFILE Age, sex, location, diversity

Macro-level analysis

- Such as policies and legislation, services, built environment, economy, social cohesion
- Feasibility of different interventions based on current and required resources, capacities and interests

Figure 8.1 IASC Analytical Framework for Durable Solutions Analysis

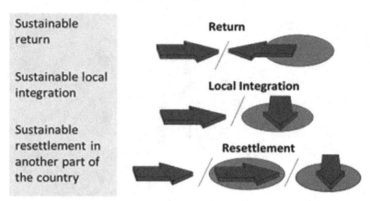

Figure 8.2 Sustainable Durable Solutions (IDMC, 2018)

for disaster risk reduction by educating the IDPs and refugees about their rights while providing inclusive, participatory mechanisms for assessing resilience in the process of decision-making for sustainable "durable solutions".

Starting with a broader understanding of the terminology of protracted displacement, qualitative data enriched the findings of this chapter towards defining the dimensions of time and space in protracted displacement in order to fill the gap in monitoring the human mobility of climate security displaced (CSD) people in urban settings. One of the main findings in this chapter is that disaggregated displacement data, generated by global organisations and authoritative bodies such as the International Organisation of Migration (IOM) and the Internal Displacement Monitoring Centre (IDMC), does not factor directly into the calculation of the number of IDPs, but it can be considered as a proxy for detailed data collection practices.

Thus, further research and advanced mechanisms are recommended to fill the gap in data sets for local integration and resettlement, and to guide the development of Urban Resilience Action Plans (U-RAPs) in fragile settings, integrating the complexities of protracted displacement into the strategies of international donors to monitor data about losses caused by disaster, action plans for city resilience, and DRR policies from a perspective of climate change and human security. Financing and long-term strategies for displaced people are needed for some learning organisations to understand risk and identify building resilience by engaging in city learning networks to integrate the context of the physical vulnerability of climate security displaced (CSD) people into the spatial planning systems of cities for DRR, "leaving no one behind". Exploring the phenomenon of protracted displacement and transformation of camps from temporary shelters into permanent settlements must be considered while developing indicators for assessing resilience in fragile settings.

Reference list

Adoko, J., & Levine, S., 2004. *Land matters in displacement: The importance of land rights in Acholiland and what threatens them.* Kampala: Civil Society Organisations for Peace in Northern Uganda.

Baytiyeh, H. 2017. Socio-cultural characteristics: The missing factor in disaster risk reduction strategy in sectarian divided societies. *International Journal of Disaster Risk Reduction,* 21: 63–69.

Beatley, T. 2014. Planning for resilient coastal communities: Emerging practice and future directions. In: *Adapting to Climate Change: Lessons from Natural Hazards Planning,* pp. 123–144. Environmental Hazards book series (ENHA). Springer Link.

Benson, C., Gyanwaly, R. P., & Regmi, H. P. 2009. *Economic and Financial Decision Making in Disaster Risk Reduction: Nepal Case Study.* Government of Nepal, Ministry of Home Affairs.

Benson, C., Twigg, J., & Rossetto, T. 2007. *Tools for Mainstreaming Disaster Risk Reduction: Guidance Notes for Development Organisations.* ProVention Consortium.

Council, N. R. & Grid, I. D. M. C. 2016. *Global report on internal displacement.* Available Online: http://www. internal-displacement. org/publications/2016/2016-global-report-on-internal-displacement-grid-2016

Expert from the Internal Displacement Monitoring Centre. 2017. (Interview, 24 July).

Expert from the International Federation of Red Cross and Red Crescent Societies. 2018. (Interview, 13 Nov).

Expert from the International Organization for Migration 2009. (Interview, 17 Nov).

Expert from the Overseas Development Institute. 2018. (Interview, 16 April).

Gall, M. 2013. *From Social Vulnerability to Resilience: Measuring Progress Toward Disaster Risk Reduction.* Unu-Ehs.

Green, L. C. 1951. United Nations General Assembly, 1950. Int'l LQ, 4, p. 216.

Habitat, U. N. 2016. *New urban agenda. Quito declaration on sustainable cities and human settlements for all.* Quito UN Habitat, pp. 1–27. Available Online: http://habitat3.org/the-new-urban-agenda/

Hynie, M. 2018. Refugee integration: Research and policy. *Peace and Conflict: Journal of Peace Psychology,* 24(3): 265.

Internal Displacement Monitoring Centre (IDMC). 2014. Internal Displacement. Global Overview of Trends and Developments in 2014. Available Online: https://www.internal-displacement.org/publications/global-estimates-2014-people-displaced-by-disasters

Internal Displacement Monitoring Centre (IDMC). 2008. *Global Overview of Trends and Developments in 2008.* Norwegian Refugee Council.

Internal Displacement Monitoring Centre (IDMC). 2018. Off the GRID. Making progress in reducing internal displacement. Norwegian Refugee Council. Available Online: https://www.internal-displacement.org/global-report/grid2018/downloads/2018-GRID.pdf

Ionescu, C., Klein, R. J. T., Hinkel, J. K. S., Kumar, K., & Klein, R. 2009. Towards a formal framework of vulnerability to climate change. *Environmental Modeling & Assessment,* 14: 1–16.

Kivela, L., Caterina, M., Elmi, K., & Lundkvist-Houndoumadi, M. 2018. *Durable Solutions Analysis Guide: A Tool to Measure Progress Towards Durable Solutions for IDPs.* Available Online: https://inform-durablesolutions-idp.org/wp-content/uploads/2018/01/Interagency-Durable-Solutions-Analysis-Guide-March2020-1.pdf

Madbouly, M. 2009. *Revisiting Urban Planning in the Middle East North Africa Region.* Regional Study Prepared for UN-Habitat Global Report on Human Settlements.

Mitchell, D., & Garibay, A. 2011. *Assessing and Responding to Land Tenure Issues in Disaster Risk Management.* Food and Agriculture Organisation of the United Nations (FAO).

OCHA. 2018. *Internal Displacement. United Nations Office for the Coordination of Humanitarian Affairs.* Available Online: https://www.unocha.org/es/themes/internal-displacement

Sanyal, R. 2017. A no-camp policy: Interrogating informal settlements in Lebanon. *Geoforum*, 84: 117–125.

Twigg, J. 2009. *Characteristics of a Disaster-Resilient Community: A Guidance Note (version 2).* Available Online: https://discovery.ucl.ac.uk/id/eprint/1346086/

United Nations. General Assembly, 1949. *Universal declaration of human rights* (Vol. 3381). Department of State, United States of America.

United Nations Human Rights Council (UNHRC). 1998. *Guiding Principles on Internal Displacement.* Available at: www.unhcr.org/protection/idps/43ce1cff2/guiding-principles-internal-displacement.html

United Nations Office for Disaster Risk Reduction. 2010. International day for disaster reduction – Aleppo, Syria. *PreventionWeb.* Available at: www.preventionweb.net/go/15980

United Nations Office for Disaster Risk Reduction (UNDRR). 2017. *Disaster Resilience Scorecard for Cities. A Tool for Disaster Resilience Planning.* Available Online: https://www.undrr.org/publication/disaster-resilience-scorecard-cities

UNHCR, UNICEF & WFP, 2018. *Vulnerability assessment of Syrian refugees in Lebanon.* Available Online: https://ialebanon.unhcr.org/vasyr/files/vasyr_reports/vasyr-2018.pdf

Waas, T., Hugé, J., Block, T., Wright, T., Benitez-Capistros, F., & Verbruggen, A. 2014. Sustainability assessment and indicators: Tools in a decision-making strategy for sustainable development. *Sustainability*, 6(9): 5512–5534.

Yassin, N. 2012. Beirut. *Cities*, 29(1): 64–73.

Zou, L., & Thomalla, F. 2008. *The causes of social vulnerability to coastal hazards in Southeast Asia.* Stockholm Environment Institute, Stockholm. Available Online: https://mediamanager.sei.org/documents/Publications/Sustainable-livelihoods/social_vulnerability_coastal_hazards_thomalla.pdf

9 Way Forward – From COP27 to COP28

The 27th Session of the Conference of the Parties (COP27) – Historical Overview of Global Targets and Commitments

The Conference of Parties (COP) is the supreme decision-making body of the United Nations Framework Convention on Climate Change (UNFCCC), which came into effect on 21 March 1994 and was ratified by 198 countries known as the "Parties to the Convention" who meet annually to review progress in dealing with climate change. The 27th Conference of the Parties (COP27) of the United Nations Framework Convention on Climate Change (UNFCCC) took place in October 2022 in Sharm el-Sheikh, Egypt. This came after the 26th session that took place in October 2021 in Glasgow, UK, hosted by the Scottish Government in partnership with Italy, to accelerate action towards the goals of the Paris Agreement and the UN Framework Convention on Climate Change. The conference was a continuation of the global action to combat climate change that was first initiated in 1979 at the first World Climate Conference (WCC).

Convened by the World Meteorological Organisation (WMO) in collaboration with the United Nations Educational, Scientific and Cultural Organization (UNESCO), the Food and Agriculture Organisation of the United Nations (FAO), the World Health Organisation (WHO), the United Nations Environment Programme (UNEP), the conference resulted in the first international recognition of greenhouse warming in response to Post-World War II global advances in basic atmospheric science and the prominence of environmental issues on the global agenda (World Meteorological Organisation, 2009). In recognition of the 1972 UN Scientific Conference First Earth Summit, a proposal was made to establish a series of stations to monitor long-term trends in the atmospheric constituents and properties and to help to guide the decision-making process for environmental protection.

In 1988, the inter-governmental Panel on Climate Change was established at the time when global warming and the depletion of the ozone layer became increasingly prominent in the international public debate and political agenda following concerns raised by the scientific community about limiting the

DOI: 10.4324/9781003363224-9

production and use of chlorofluorocarbons F-11 and F-12, which led to the adoption of the Vienna Convention for the Protection of the Ozone Layer, the ratification of the 1987 Montreal Protocol, and the conclusion of a protocol from the 1979 Transboundary Air Pollution Convention. The Montreal Protocol is considered one of the most successful, multi-lateral, environmental treaties in modern history, as it resulted in the elimination of nearly 99% of chlorofluorocarbons (CFCs) and ozone-depleting substances, and was amended in the 2016 Kigali Agreement for countries to reduce their production of hydrofluorocarbons (HFCs). Nevertheless, further investigation is required to understand: How and what were the mechanisms for committing member states? Who ratified the treaty? What actions to monitor impact took place? How can a similar approach be applied currently to enhance the commitments of national and local governments?

A number of factors contributed to the success of the Montreal Protocol, including a growing public concern about ozone depletion and strong scientific evidence of the impacts it has on the environment and health, as well as advocacy of the development and adoption of alternative technologies that do not deplete the ozone. Developing countries were also provided with financial and technical assistance for phasing out ozone-depleting substances. This was one of the key elements that resulted in the success of the protocol that should be adopted to help to achieve the 2023 global targets.

The protocol also provided for financial and technical assistance to developing countries to help them to phase out the use of ozone-depleting substances. The treaty requires countries to agree formally to be bound by its provisions. In ratifying the Montreal Protocol, the European Union and all member states of the United Nations demonstrated their commitment and established a mechanism to monitor the impact of the agreement, to report regularly on its implementation, and to provide information on the production and consumption of ozone-depleting substances.

The First Assessment Report of the Inter-governmental Panel on Climate Change (IPCC) shaped the agenda of the 1990 Second World Climate Conference (WCC-2), which involved 747 participants from 116 countries. With the aim of reviewing the first decade of progress under the WCC-1, this was followed by the first meeting of the Inter-governmental Negotiating Committee (INC) for a UN Framework Convention on Climate Change (UNFCCC) in 1991 and the signing of the WCC-1 at the Rio Earth Summit in 1992 (Boyle, 1994). With specific commitments to systematic observation and research in support of its ultimate objective of "stabilisation of greenhouse-gas concentrations in the atmosphere at a level that would prevent dangerous anthropogenic interference with the climate system", the UNFCCC was ratified by 197 countries, signed by 155 countries, and came into effect in 1994.

Adopting the success of the Montreal Protocol in binding member states to act in the interests of human safety even in the face of scientific uncertainty, the UNFCCC established an annual forum under the title Conference of the

Parties (COP), which acts, to date, as a global platform for international discussions aimed at stabilising the concentration of greenhouse gases in the
atmosphere. This has led to the establishment of key milestones in climate
negotiations, such as the Kyoto Protocol and the Paris Agreement, as noted in
Chapter 1 (United Nations, 2007).

Considered to be two landmark agreements in the multi-lateral climate
change process, both took the form of legally-binding, international treaties
to limit global warming. The aim of the 2015 Paris Agreement was to prevent average global temperatures reaching above 1.5°C and was adopted by
196 parties at COP21 in Paris. Operating within a five-year cycle, countries
submitted their plans in 2020 for climate action, known as Nationally Determined Contributions (NDCs), and were encouraged to communicate actions
that they take to reduce their greenhouse gas emissions through long-term
low greenhouse gas emission development strategies LT-LEDS. With limited
emphasis on mechanisms for adaptation to climate change, countries continue
to update their NDCs but fall behind in producing the $100 billion funding for
developing nations, promised 11 years ago at COP16, and finalising the rules
of the Paris Climate Accord adopted in 2015. With emphasis on environmental protection, direct inter-linkages with human development were not established until the launch of the 2015–2030 Sustainable Development Goals.

In 2021, COP26 concluded with delivering the Glasgow Climate Pact, the
aim of which is to accelerate action on climate during this decade. By then,
reports on NDCs driving global actions for mitigation, adaptation, financing,
and collaboration were submitted from 153 countries, a new UN climate programme on mitigation ambition was launched, and the Paris Rulebook was
finalised, which is a new reporting mechanism and standardised transparency
framework developed for international carbon markets in which improved and
common timeframes for reaching targets for reductions in emissions are called
for, urging the world's largest carbon emitters for commitments to move away
from coal power and reduce their methane emissions with further actions to be
taken to protect natural bio-diversity and halting and reversing deforestation.

At COP27, national governments agreed to establish a "transitional committee" to make recommendations on how to operationalise the Santiago
Network for Loss and Damage, and the Glasgow–Sharm el-Sheikh Work Programme for Progress on the Global Goal for Adaptation, which will conclude
at COP28. These efforts will help to inform the first Global Stocktake and
improve climate risk governance amongst the most vulnerable. However, this
cannot be achieved without effective, multi-level governance to transform cities into low-emission, resilient urban systems for a better urban future for all.
Reinforcing multi-level co-operation to direct funds and technical assistance
for sub-national governments is critical to collect and monitor reductions of
carbon emission in cities and regions and to help to achieve global targets.

With the impact of climate change evident mostly in least developed countries
and small island states, calls to boost global funds to increase preparedness for

climate risks were raised, resulting in the development of the Glasgow–Sharm el-Sheikh Work Programme for Progress on the Global Goal on Adaptation, bridging the gap between COP26 and COP27 while preparing the way for more smooth negotiations on pledges to double the 2019 levels of finance for adaptation by 2025 and achieving the goal of $100 billion in finance for CCA by 2023. Following up on the COP19 and the Warsaw International Mechanism for Loss and Damage, the Santiago Network on Loss and Damage was brought into the discussions to improve access to financing for adaptation and to support the launch of the Least Developed Countries Fund.

MENA Region – Climate Change Mitigation, Adaptation, and Financing

Hosting the world's most water-stressed countries, the MENA Region is under growing pressure of water scarcity and food insecurity as a result of shrinking freshwater supplies, reduced agriculture yields, and advancing desertification because 75% of agriculture in the region depends on rainfall, which is expected to decline in the coming decades. Rising sea level and coastal erosion is another major hazard caused by an increase of 1.5°C in global warming, affecting more than 60% of the region's population living in coastal cities, with expectations of a rise in sea level by 0.6m on average in a +4°C world and for Maghreb to experience a 1.2m rise in sea-level by 2080 (WB, 2022).

With an estimate of 20 million internal climate migrants in North Africa as a result of rising sea levels (WB, 2022), it is critical to integrate the vulnerabilities and needs of climate security displaced (CSD) people into the national strategies of countries for climate mitigation and adaptation.

Oil-producing countries must address inefficiencies in their extractive activities to lead the green transition to low-carbon economies and reduce dependence on fossil fuels (constituting 50% of exports on average), advance energy efficiency, and promote investments in renewable energy.

At COP27, a specific fund for loss and damage responding to the recent figures on economic losses from countries reporting on SFDRR- Target C indicated that, between 2015 to 2021, the annual reported losses reached an average of USD 330 billion, which represented a full 1% of countries' global GDP (UNDRR & WMO, 2022). However, only the limited number of 24 out of 193 nations submitted updated plans on Nationally Determined Contributions (NDCs). Thus, there is a lack of integration in the initiatives or commitments announced in Glasgow, and delays in the delivery from developing countries of the 2020 annual target of $100 billion in finance for adaptation to climate (United Nations, 2021). Despite the low number of updated NDCs, approximately 84% have strong or moderate urban content compared with 69% at COP26. This indicates the significant role of national governments in accelerating subnational action regarding climate and in providing the enabling environment for the inclusive participation of all stakeholders within the urban system. This is critical to achieve new commitments regarding climate and to help to prioritise key issues associated with displacement caused by climate, such as tenure security and housing accessibility and affordability, to be introduced in the forefront of national adaptation plans and to be integrated into urban planning strategies of cities to advance public investments in local mechanisms for climate adaptation and mitigation.

In addressing the region's targets for mitigation of climate change, it is important to consider the expected increase in demand for energy to double by 2030 with the production of oil projected to increase from 35% to 44% in order to advance decarbonisation efforts and reduce transition risks. The challenges of modelling macro-economic outcomes of physical and transition risks must be monitored to help to translate the impacts of transition risks into actions for resilience and to define macro-economic and sectoral outcomes that are critical to reduce exposure and vulnerability to climate change (Green, 2021).

"Unlike mitigation of GHG emissions, climate adaptation does not have a universal metric and its ambition or implementation level cannot simply be aggregated based on countries' national pledges" (Leiter, 2022). Thus, the three dimensions of the Adaptation Gap Report of the United Nations Environment Programme (UNEP) will be investigated together with the UNDRR Disaster Resilience Scorecard Ten Essentials to help to monitor progress on action for adaptation, and to establish a standardised mechanism to develop Urban Resilience Action Plans (U-RAPs) that can be applied in the MENA Regional context and beyond while addressing its fragility and vulnerabilities under the UNDP key themes of planning, financing, and implementing actions for adaptation (Table 9.1).

To address the increasing demands to improve financing for CCA, new pledges were made at COP27, such as the Adaptation Fund, totalling more than USD 230 million, the G7 Global Shield Financing Facility and the World Bank partnership, under the name SCALE – Scaling Climate Action by Lowering Emissions, the aim of which is to help to catalyse transformative climate action.

Yet, without advancing and directing CCA investments towards building capacities of both the public and private sector, these efforts cannot address energy poverty and unlock the financing for urban adaptation to climate in fair and just plans for decarbonisation and net-zero transition. Oil producing countries in the MENA Region, such as Saudi Arabia and the UAE, can play a major role in advancing transformational, alternative technologies such as green hydrogen. "UAE could capture 25% of the low-carbon hydrogen market, the value of which is expected to reach more than $400 billion per year over the next five years" (WB, 2022).

Associated with climate justice, at the first-ever Ministerial Meeting on Urbanisation and Climate Change that took place at COP27, the importance of local and regional action in encouraging commitments to climate was highlighted (UN HABITAT, 2022). Convened as part of Solutions Day, it was an

Table 9.1 Mapping UNEP Adaptation Principles and Disaster Resilience Scorecard Essentials

UNEP Adaptation Gap	Key Components of Urban Resilience Action Plan (U-RAP)				
	Planning		*Financing*	*Implementation*	
UNDRR New Ten Essentials	Sensitisation put in place for local DRR organisational structure with links to city vision and strategy	Technical analysis to identify the city's risks and vulnerabilities in targeted areas	Provide financing options to evaluate alternative investment mechanisms	Enhance investments in building institutional and technical capacities	Project implementation and prioritisation of actions	Monitoring system and evaluation mechanism to assess progress
Essential 1	Organise for disaster resilience					
Essential 2		Identify, understand, and use current and future risk scenarios				
Essential 3			Strengthen financial capacity for resilience			
Essential 4					Pursue resilient urban development and design	

Essential 5

Safeguard natural buffers to enhance the protective functions offered by natural ecosystems

Essential 6

Strengthen institutional capacity for resilience

Essential 7
Essential 8

Increase infrastructure resilience

Essential 9

Ensure effective preparedness and disaster response

Essential 10

Expedite recovery and build back better

important step in achieving alignment of the multi-layered actions regarding climate, where ministries of housing, urban development, environment, and climate change from 50 member states were present, as well as representatives from local and regional governments, civil society, non-governmental organisations, urban networks, multi-lateral banks, UN institutions, and inter-governmental organisations. With the presence of country officials from Mauritania, Morocco, Tunisia, Sudan, Algeria, Libya, and Palestine, national governments in the MENA Region should play a stronger role in accelerating sub-national action regarding climate and providing an enabling environment for the inclusive participation of all stakeholders within the urban system. This will help to achieve new climate commitments with financing for urban resilience being integrated into strategies for urban planning of cities and public investments in adaptation and mitigation.

Policy Recommendations and Actions – COP28

Despite global efforts of climate governance being directed towards mitigation of greenhouse gas (GHG) emissions, adaptation was announced as a key priority for both Presidencies of the 26th and 27th Conferences of the Parties (COP) to the United Nations Framework Convention on Climate Change (UNFCCC). In order to establish resilient, low-carbon societies and promote equitable development, peace, and stability in the Region's most fragile settings, the nations of the Middle East and North Africa (MENA) must expedite system-wide green transitions and create long-term approaches to adaptation to climate. The World Bank Middle East and North Africa Climate Roadmap 2021–2025 indicates "adopting an economy-wide, whole of government approach to strengthen institutions, overcome barriers to private sector engagement, foster regional integration – while ensuring that no one is left behind" (WBb, 2022).

Societal Resilience is key for developing an inclusive and sustainable Policy Guidance for Urban Resilience and Climate Adaptation to address the "social fabric of the place" while differentiating between adaptive, absorptive and transformative capacities of "community" and "society" as part of the city-wide system and institutional structure. Moving forward from disaster management to disaster risk reduction requires measuring and monitoring the impact of climate change in developing and implementing an Urban Resilience Action Plan (U-RAP) in the short- and long-term strategy of city development and durable solutions for climate security displaced (CSD) people. Considering the complexity of the context of climate security, conflict and displacement in the Middle East and North Africa Region and beyond, this book contributes to existing knowledge by framing evidence-based guidance for improving the overall regulatory environment for risk governance. Integrating sustainability into the decision-making process for DRR protection and prevention will enable the legislation of the rights of IDPs and refugees in pre- and post-disaster stages.

Taking into account the challenges of urban governance in the MENA Region concerning the lack of transparency and limitations in financial and human capacities, it is affirmed in this chapter that biases in socio-economic assessment might remain as the main obstacle to the application of Open Data in the decision-making process for assessing resilience. It is important to improve human and technical capacities with the use of Open Data to develop a comprehensive action plan for urban resilience and to obtain consistent reporting on climate security displacement and losses of data for all hazards and underlying risks caused by disaster. Further research can be developed from the findings of the book to create evidence-based records on the implementation of SFDRR beyond 2020, and the level of achieving Target- E: Substantially increase the number of countries with national and local disaster risk reduction strategies at local, national, regional, and global levels.

To assess progress from COP27 to COP28 on Climate Change Adaptation (CCA) it would be important to consider the commitments made by countries in their nationally determined contributions (NDCs) and other action plans to address climate. These commitments should include measurable targets for adaptation such as increased investment in resilient infrastructure, early warning systems for extreme weather events, and measures to protect vulnerable communities and ecosystems. Leiter (2022) attempted to provide an overview of how negotiations about adaptation evolved in the first three years following the adoption of the Paris Agreement. Indicating that "understanding of negotiation outcomes requires observing the negotiation process first-hand, rather than interpreting final decision texts in the absence of knowing how they evolved", Leiter (2022) outlined the soft law characteristic of the Paris Agreement with its low obligations and low precision regarding adaptation. This argument is supported by the IPCC Sixth Assessment Report in which it was highlighted that "most observed adaptation is fragmented, small in scale, incremental, sector-specific, designed to respond to current impacts or near term risks, and focused more on planning rather than implementation" (IPCC, 2011). This is strongly associated with the reality that adaptation also requires anticipation, especially in sectors with long-term investments (Hallegatte *et al.*, 2008).

With the aim of helping to achieve COP27 commitments on all scales, this book will benefit two main beneficiaries in direct and indirect courses of action. The direct beneficiaries are the decision-makers in building Disaster Risk Resilience in the Arab Region by providing policy guidance on integrating indicators of adaptation into the assessment of urban resilience while helping to support the efforts of national and local governments towards identifying local techniques to mitigate the impacts of climate change and human insecurity, and to guide financial investments towards the generation of better data about loss and damage at the local level while enhancing monitoring and early warning on small-scale onset and frequent hazardous events that are not registered in international databases about loss caused by disaster associated with urban mobility patterns and displacement caused by conflict in fragile settings (Bharadwaj and Shakya, 2021).

The benefit of this publication will extend to formulating guidelines on how to inform the process of assessing resilience and measuring the vulnerabilities of climate security displaced (CSD) people as indirect beneficiaries beyond emergency humanitarian needs of shelter and food, to include sustainable development pillars of tenure security, land ownership, access to education, employment, public services, and critical infrastructure.

Addressing the key question of how to ensure that the wider financing system is not acting against but directing capital towards building resilience, it is necessary to understand the underlying conditions of macro-economics that hinder private and public investments in transitions to low carbon and adaptation to climate change in cities and regions. Cities and regions have limited means, capacity, and financial resources with increasing public debts because of high inflation rates and increasing severity and frequency of climate-related disasters.

Any global adaptation programme or goal can only be achieved through local and regional adaptation and resilience action. It is necessary to integrate considerations of resilience to climate (physical and transitional risks) into financial decision-making and address the gap in the lack of adequate information on climate risks and vulnerabilities, to help to measure the contributions of financing activities to resilience to climate, and to assess and disclose climate risks and opportunities for resilience to climate in business and financing operations.

With financing needs estimated at $186 billion in the nationally determined contributions (WB, 2022), providing a common language that can be used across a diverse range of financial institutions, role-players (banks, pension funds, insurance companies, corporations, impact investors, and other private agencies) will contribute to providing debt finance for adaptation and financing operations and help to enhance economic competitiveness, enhance financial stability, social cohesion, and co-benefits of climate policies (Hallegatte and Dumas, 2008). It is important for challenging the rationale for stand-alone adaptation projects to be associated with "resilience upgrading" to set financially justified risk reduction measures and increase the reliability of investment returns and asset values while recognising the variations in the financing operations in the context of the MENA Region that require different approaches and will lead to better integration of Climate Change Adaptation (CCA) operations and reduction in GHG emissions.

Reference list

Bégni, G., Darras, S., & Belward, A., 2021. The Kyoto Protocol: Legal statements, associated phenomena and potential impacts. In *Observing Our Environment from Space-New Solutions for a New Millennium* (pp. 9–21). CRC Press.

Boyle, A. E. 1994. Negotiating climate change: The inside story of the Rio Convention. Mintzer, I. M., & Leonard, J. A. (Eds). Paper, pp. xiv, 392, Index, 22.95. Cambridge, New York: Cambridge University Press. *American Journal of International Law*, 89(4): 864–865.

Green, J. F. 2021. How Reforming Tax and Trade Rules Can Fight Climate Change. *Follow the Money*. Available at: www.foreignaffairs.com/articles/world/2021-11-12/follow-money

Hallegatte, S. 2008. An adaptive regional input-output model and its application to the assessment of the economic cost of Katrina. *Risk Analysis: An International Journal*, 28(3): 779–799.

Hallegatte, S., & Dumas, P. 2008. *Adaptation to Climate Change: Soft vs. Hard Adaptation*. OECD Expert Workshop on Economic Aspects of Adaptation to Climate Change.

IPCC. 2011. Climate change mitigation. Special report on renewable energy sources and climate change mitigation. *Renewable Energy*, 20(11).

Leiter, T. 2022. Too little, too slow? Climate adaptation at the United Nations climate change negotiations since the adoption of the Paris Agreement. *Carbon and Climate Law Review*, 16(4).

UN HABITAT. 2022. Ministerial meeting on urbanization and climate change. *COP27 Solutions Day*. Available at: https://unhabitat.org/sites/default/files/2023/04/cop27_ministerial_meeting_on_urbanization_and_climate_change_final.pdf

UNDRR and WMO, 2022. Global Status of Multi-Hazard Early Warning Systems: Target G, Geneva, United Nations Office for Disaster Risk Reduction, World Meteorological Organization. Available Online: https://www.undrr.org/publication/global-status-multi-hazard-early-warning-systems-target-g

United Nations. 2007. *From Stockholm to Kyoto: A Brief History of Climate Change*. Available at: www.un.org/en/chronicle/article/stockholm-kyoto-brief-history-climate-change

United Nations. 2021. *Delivering the Glasgow Climate Pack*. Uniting the World to Tackle Climate Change. Available at: https://ukcop26.org/

United Nations. *History of the Convention*. Available at: https://unfccc.int/process-and-meetings/what-is-the-united-nations-framework-convention-on-climate-change

World Bank. 2022a. *COPs Offer Middle East-North Africa a Climate Leadership Role*. Available at: www.worldbank.org/en/news/opinion/2022/11/16/cops-offer-middle-east-north-africa-a-climate-leadership-role

World Bank. 2022b. *Middle East and North Africa Roadmap (2021–2025)*. Driving Transformational Climate Action and Green Recovery in MENA.

World Meteorological Organisation. 2009. *A History of Climate Activities*. Available at: https://public.wmo.int/en/bulletin/history-climate-activities

Index

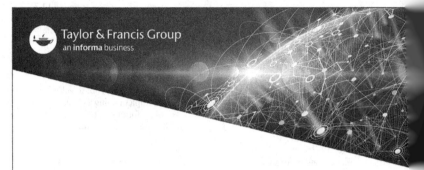

Printed in the United States
by Baker & Taylor Publisher Services

Printed in the United States
by Baker & Taylor Publisher Services